Exoplanet Atmospheres

PRINCETON SERIES IN ASTROPHYSICS

Edited by David N. Spergel

Theory of Rotating Stars, *by Jean-Louis Tassoul*

Theory of Stellar Pulsation, *by John P. Cox*

Galatic Dynamics, *by James Binney and Scott Tremaine*

Dynamical Evolution of Globular Clusters, *by Lyman Spitzer, Jr.*

Supernovae and Nucleosynthesis: An Investigation of the History of Matter, from the Big Bang to the Present, *by David Arnett*

Unsolved Problems in Astrophysics, *edited by John N. Bahcall and Jeremiah P. Ostriker*

Galactic Astronomy, *by James Binney and Michael Merrifield*

Active Galactic Nuclei: From the Central Black Hole to the Galactic Environment, *by Julian H. Krolik*

Plasma Physics for Astrophysics, *by Russell M. Kulsrud*

Electromagnetic Processes, *by Robert J. Gould*

Conversations on Electric and Magnetic Fields in the Cosmos, *by Eugene N. Parker*

High-Energy Astrophysics, *by Fulvio Melia*

Stellar Spectra Classification, *by Richard O. Gray and Christopher J. Corbally*

Exoplanet Atmospheres: Physical Processes, *by Sara Seager*

Exoplanet Atmospheres

Physical Processes

Sara Seager

PRINCETON UNIVERSITY PRESS
PRINCETON AND OXFORD

Published by Princeton University Press, 41 William Street,
Princeton, New Jersey 08540

In the United Kingdom: Princeton University Press, 6 Oxford Street,
Woodstock, Oxfordshire OX20 1TW

ISBN: 978-0-691-11914-4 (cloth)
ISBN: 978-0-691-14645-4 (pbk.)

Library of Congress Control Number: 2010924851

British Library Cataloging-in-Publication Data is available

The publisher would like to acknowledge the author of this
volume for providing the camera-ready copy from which this
book was printed.

This book has been composed in Times and Helvetica.

Printed on acid-free paper ∞
press.princeton.edu
Printed in the United States of America

1 3 5 7 9 10 8 6 4 2

Dedicated to the memory of my mentor

Dr. John N. Bahcall

Contents

Preface

A new era in planetary science is upon us. Hundreds of extrasolar planets ("exoplanets") orbiting normal stars are known. Surprisingly, the exoplanetary systems are very different from our own solar system. Gas giant planets have been detected at a wide range of orbital distances from the parent star, including some much closer to their parent star than Mercury is to our Sun. Many exoplanets have eccentric orbits, some with eccentricities up to 0.9. Planets with no solar system analogs are being discovered including mini Neptunes and super Earths. Now that the existence of exoplanets is firmly established, the adventure of exploring their physical characteristics has begun in earnest. An exoplanet's physical properties, such as density, atmospheric composition, and atmospheric temperature, can be measured for a subset of exoplanets.

The first fifteen years of exoplanet discoveries have taught us to expect surprises, because the random nature of planet formation and migration leads to many different planetary system architectures. These include an astonishing range of observed exoplanet masses, semimajor axes, and orbital eccentricities. We similarly anticipate a huge diversity of exoplanet atmospheres; the formation and composition of an atmosphere will depend on where in the disk a planet forms, its evolutionary history (due in part to atmospheric escape), and its present semimajor axis.

To understand the physical characteristics of the potentially wide variety of planetary atmospheres, knowledge of the general physics common to all planetary atmospheres—both exoplanets and solar system planets—is needed. The goal of this book is to present the basic physics that can be used to obtain fundamental atmospheric characteristics and the first-order vertical thermal structure from spatially unresolved spectra and photometry. Emphasis will be on the major physical processes that govern the planetary atmosphere and emergent spectrum. This book is aimed at advanced undergraduate students, graduate students, and researchers entering the field of exoplanet atmosphere studies.

With hundreds of exoplanets now known, and dozens of exoplanet atmosphere observations, there is a need for a book that addresses, in a general way, the basic planetary atmospheric physics for planetary atmospheres in a broad range of environments. Although the physics is the same as for solar system planet atmospheres, traditional planetary atmosphere books are necessarily specific and descriptive in addressing highly detailed observations and phenomenology of solar system planets. While textbooks on stellar atmospheres address many of the issues common to exoplanet atmospheres, they lack topics highly relevant for a nonluminous planet orbiting a star (such as transmission spectra, scattered and polarized radiation, and thermal orbital phase curves). A basic understanding of atmospheric processes is necessary to interpret data and to aid experiment design.

This is a tremendously exciting time for exoplanet atmosphere studies. Dozens of exoplanet atmospheres have been observed, mostly from space. So far, atmospheric observations and interpretation have been limited to a subset of exoplanets: hot Jupiters or hot Neptunes in excruciatingly close orbits to their host stars; and directly imaged young, massive exoplanets in very distant orbits around their host star. This will all change in the not-too-distant future. The anticipated launch of NASA's *James Webb Space Telescope* in 2014 as well as new ground-based instrumentation will enable a wider range of planets (in terms of masses and orbits) whose atmospheres can be studied. In the more distant future, we hold hope for a space telescope capable of direct imaging, for both discovery and follow-up atmosphere measurements of true Earth analogs.

Good luck on your journey to explore exoplanet atmospheres!

Many people deserve thanks for help with this book. First and foremost a very special thanks to my husband Mike Wevrick, for help with figures, the index, editing, and grammar. His continued patience in many aspects helped tremendously. Second I thank Cindi Fluekiger for her support and encouragement. Many students and colleagues helped with the description of complicated concepts or read drafts of the book for content and accuracy. Thanks to Bjoern Benneke, Renyu Hu, Hannah Jang-Condell, Jim Kasting, Mark Marley, Paul O'Gorman, Dimitar Sasselov, Feng Tian, Maggie Turnbull, and Neil Zimmerman. A special thanks to Nikku Madhusudhan and Leslie Rogers for their attention to detail. For technical assistance I thank Meghan Kanabay and Eugene Jang for help with figures and Lucy Day Werts Hobor and Steve Peter for their exceptional LaTeX expertise. Finally I express my thanks to the Princeton University Press staff Ingrid Gnerlich, Brigitte Pelner, and Dimitri Karetnikov for very professional oversight of this project.

Sara Seager
Cambridge, MA

Exoplanet Atmospheres

Chapter One

Introduction

1.1 EXOPLANETS FROM AFAR

The search for our place in the cosmos has fascinated human beings for thousands of years. For the first time in human history we have technological capabilities that put us on the verge of answering such questions as, "Do other Earths exist?," "Are they common?," and "Do they harbor life?" Critical to inferring whether or not a planet is habitable or inhabited is an understanding of the exoplanetary atmosphere. Indeed, almost all of our information about temperatures and chemical abundances in planets comes from atmospheric photometry or spectroscopy.

Ultimately we would like an image of an Earth twin as beautiful as the Apollo images of Earth (Figure 1.1). For our generation we are instead limited to observing exoplanets as spatially unresolved, i.e., as point sources. The *Voyager I* spacecraft viewed Earth in such a way from a distance of more than four billion miles (Figure 1.1). The Earth's features are hidden in a pale blue dot's tiny speck of light.

For exoplanets, we can potentially measure brightnesses, brightness temperatures, and spectra. Earth itself has been observed "as an exoplanet." We show Earth's visible- and mid-infrared wavelength spectrum in Figure 1.2. As we shall

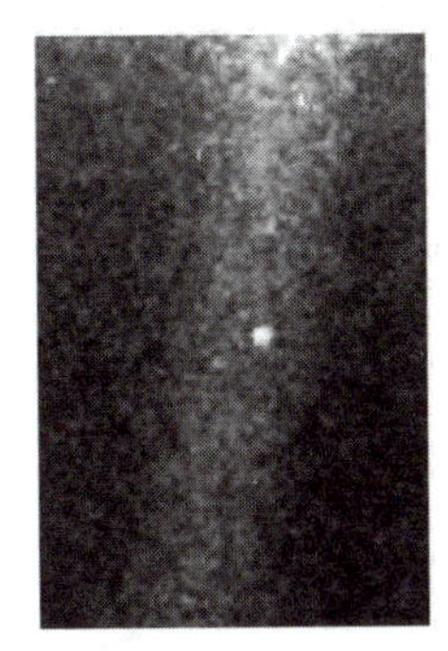

Figure 1.1 Earth as viewed from space. Left: image from NASA's *Apollo 17* spacecraft in 1972. Right: image from *Voyager I* at a distance of more than four billion miles. The Earth lies in the center of a band caused by scattered light in the camera optics.

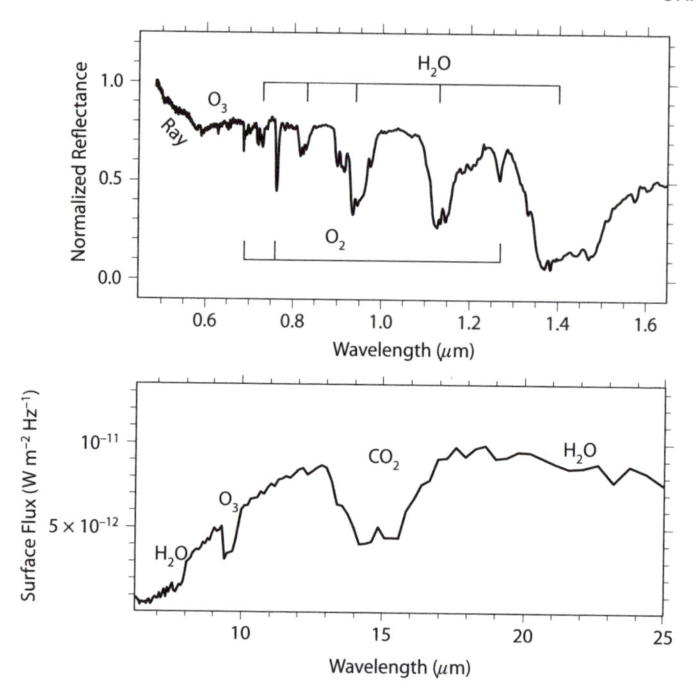

Figure 1.2 Earth's hemispherically averaged spectrum. Top: Earth's visible wavelength spectrum from Earthshine measurements [1]. Bottom: Earth's mid-infrared spectrum as observed by Mars Global Surveyor enroute to Mars [2]. Major molecular absorption features are noted; Ray means Rayleigh Scattering.

see in later chapters of this book, a spectrum contains information about a planet atmosphere's temperature and composition.

We cannot detect and study an Earth twin with current telescopes and instrumentation. Earth's pale blue dot is adjacent to our Sun, a star that is many orders of magnitude brighter than Earth. To be precise, the planet-star visible wavelength contrast for an Earth-Sun twin is 2.1×10^{-10} and for a Jupiter-Sun twin is 1.4×10^{-9}. The fundamental difficulty in observing exoplanet atmospheres is not their intrinsic faintness—indeed they are no fainter than the faintest galaxies observed by the *Hubble Space Telescope*. Instead, the enormous planet-star contrast is the major impediment to the direct observation of exoplanets.

1.2 TWO PATHS TO OBSERVING EXOPLANET ATMOSPHERES

The most natural way for us to think of observing exoplanets—albeit as point sources—is by taking an image of the exoplanet. This would be akin to taking a photograph of the stars with a digital camera, although using a very expensive, high-quality detector. This so-called direct imaging of planets is currently limited

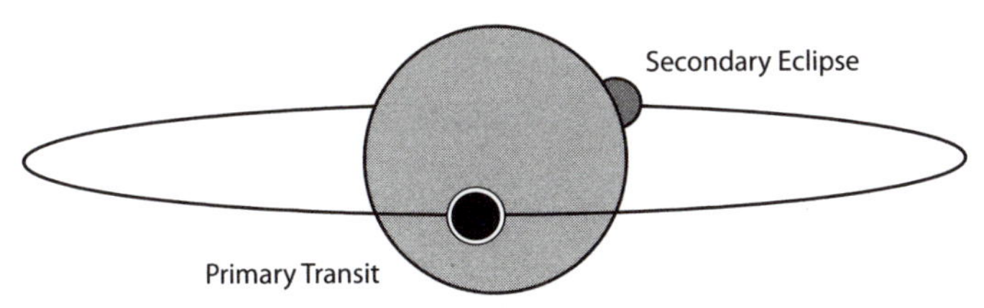

Figure 1.3 Diagram of a transiting exoplanet going in front of and behind its parent star.

to big, bright, young, or massive substellar objects located far from their stars [e.g., 3, 4]. Direct imaging of substellar objects is currently possible from space or with large ground-based telescopes and adaptive optics to cancel the atmosphere's blurring effects. For current technology, direct imaging of an Earth twin or Jupiter twin is out of the question. The high planet-star contrasts are prohibitive. Fortunately, much research and technology development is ongoing for space-based direct imaging to enable direct imaging of solar-sytem-aged Earths and Jupiters in the future.

For the present time, two fortuitous, related events have enabled observations of exoplanet atmospheres in a manner very different from direct imaging. The first event is the existence and discovery of a large population of planets orbiting very close to their host stars. These so-called hot Jupiters, hot Neptunes, and hot super Earths have less than four-day orbits and semimajor axes smaller than 0.05 AU. The hot Jupiters are heated by their parent stars to temperatures of 1000–2500 K, making their infrared brightnesss only $\sim$1/1000 that of their parent stars. While it is by no means an easy task to observe a 1:1000 planet-star contrast, this situation is unequivocally more favorable than the 10^{-10} visible-wavelength planet-star contrast for an Earth twin orbiting a sunlike star.

The second favorable occurrence is that of transiting exoplanets—planets that go in front of their star as seen from Earth. The closer the planet is to the parent star, the higher its probability to transit. Hence the existence of short-period planets has ensured that there are many transiting planets. It is this special transit configuration that allows us to observe the planet atmosphere without direct imaging (Figure 1.3).

Transiting planets are observed in the combined light of the planet and star (Figure 1.4.) As the planet goes in front of the star, the starlight drops by the amount of the planet-to-star area ratio. If the size of the star is known, the planet size can be determined. During transit, some of the starlight passes through the optically thin part of the planet atmosphere (depicted by the transparent annulus in Figure 1.3), picking up some of the spectral features in the planet atmosphere. By comparison of observations of the superimposed planet atmosphere taken during transit with observations of the star alone (outside of transit), the planet's transmission spectrum can in principle be measured.

Planets in circular orbits that go in front of the star also go behind the star. Just before the planet goes behind the star, the planet and star can be observed together. When the planet goes behind the star only the starlight is observed. These two observations may be differenced to reveal emission from the planet alone.

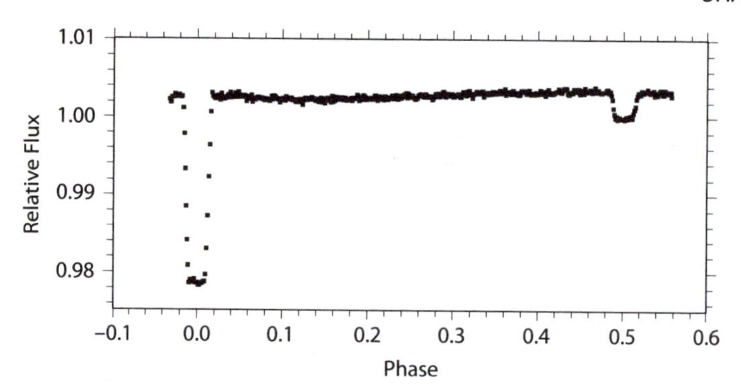

Figure 1.4 Infrared light curve of HD 189733A and b at 8 μm. The flux in this light curve is from the star and planet combined. The first dip (from left to right) is the transit and the second dip is the secondary eclipse. Error bars have been suppressed for clarity. See Figure 1.3. Data from [5].

The primary transit and secondary eclipse of the planet HD 189733b at 8 μm are shown in the combined light of the planet-star system in Figure 1.4. HD 189733 is a star slightly smaller than the Sun with a radius about 75% of the Sun's radius and the planet has an effective temperature of about 1200 K. At 8 μm we are seeing the thermal emission from the planet. As we will see in Chapter 3, the planetary brightness temperature can be measured by considering the depth of the secondary eclipse.

1.3 TYPES OF PLANETS

The most surprising aspect of the hundreds of known exoplanets is their broad diversity: a seemingly continuous range of masses and orbital parameters (Figure 1.5). Planet formation is a stochastic process, whereby the randomness likely gives birth to planets of a wide range of masses in a wide variety of locations in a protoplanetary disk. Planetary migration allows planets to end up very close to the parent star. We expect to find a diversity in exoplanet atmospheres that rivals the diversity in exoplanet masses and orbital parameters.

The solar system itself contains an interesting menagerie of planets. Despite their very different visual appearances, the solar system planets are usually divided into two main categories: terrestrial planets and giant planets. The terrestrial planets include the four inner planets, Mercury, Venus, Earth, and Mars. The terrestrial planets are predominantly composed of rock and metals and have thin atmospheres. Atmospheric evolution, from both atmospheric escape of light gases and gas-surface reactions, has substantially changed each of the terrestrial planet atmospheres from their primitive state.

The giant planets include the four outer planets, Jupiter, Saturn, Uranus, and Neptune. These planets are vastly different from the terrestrial planets, with no rocky surfaces and masses up to hundreds of times those of the terrestrial planets.

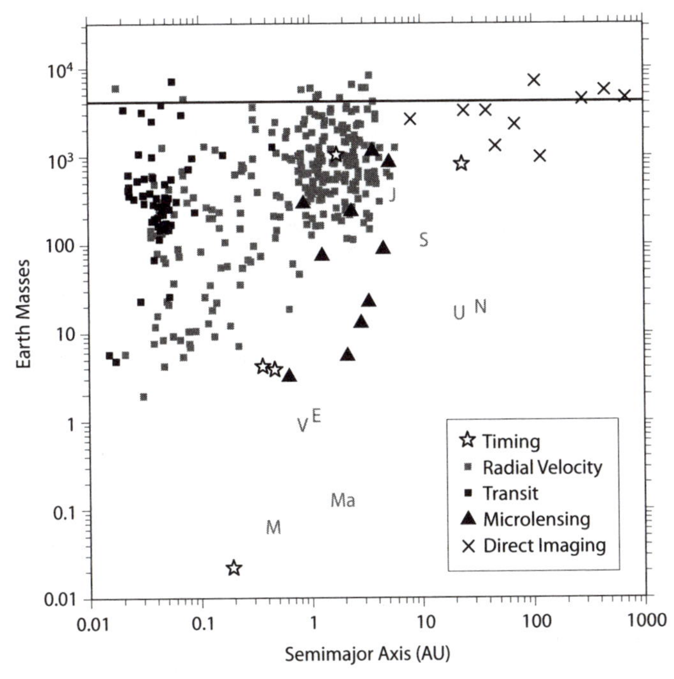

Figure 1.5 Known exoplanets as of January 2010. Exoplanets surprisingly are found at a nearly continuous range of masses and semimajor axes. Many different techniques are successful at discovering exoplanets, as indicated by the different symbols. The circles surround planets with an atmosphere measurement. The solar system planets are denoted by their first initial. The horizontal black line is the conventional upper limit to a planet mass, 13 Jupiter masses. The sloped, lower boundary to the collection of grey squares is due to a selection effect in the radial velocity technique. Data taken from [6].

The "gas giant" planets Jupiter and Saturn are composed almost entirely of hydrogen and helium with massive atmospheres and liquid metal interiors. The "ice giant" planets Uranus and Neptune are composed of mostly water, methane, and ammonia ices, yet also have a 1–2 Earth-mass envelope of hydrogen and helium. In contrast to the terrestrial planets, the giant planet atmospheres are primitive—little atmospheric evolution has taken place so that they contain roughly the same atmospheric gases as at their formation.

We turn to a detailed definition of planet types, in part taken from [7]. A planet is an object that is gravitationally bound and supported from gravitational collapse by either electron degeneracy pressure or Coulomb pressure, that is in orbit about a star, and that, during its entire history, never sustains any nuclear fusion reactions in its core. Reliance on theoretical models indicates that such objects are less massive than about 13 Jupiter masses. A lower mass limit to the class of objects called planets has not been convincingly determined. An exoplanet is a planet orbiting a star other than the sun. (We note the term exoplanet has been used to include "free floating" planets that lack a host star.)

A terrestrial planet is a planet which is primarily supported from gravitational collapse through Coulomb pressure, and which has a surface defined by the radial extent of the solid interior or liquid outer layer. Terrestrial planets are often referred to as "rocky planets". A thin gas atmosphere may exist above the surface.

A giant planet is a planet with a mass substantially greater than terrestrial planets but less than brown dwarfs, e.g., $0.03\ M_\mathrm{J} < M_\mathrm{p} < 13 M_\mathrm{J}$ ($\sim 10\ M_\oplus < M_\mathrm{p} \lesssim 4000\ M_\oplus$). Giant planet interiors are thought to contain a substantial fraction of metallic hydrogen surrounding a rocky core.

A potentially habitable planet is one whose orbit lies within the star's habitable zone. The habitable zone is the region around a star in which a planet may maintain liquid water on its surface. The habitable zone boundaries can be defined empirically based on the obervation that Venus appears to have lost its water some time ago and that Mars appears to have ahd surface water early in its history. The habitable zone can also be defined with models. A habitable planet is a terrestrial planet on whose surface liquid water can exist in steady state. This definition presumes that extraterrestrial life, like Earth life, requires liquid water for its existence. Both the liquid water, and any life that depends on it, must be at the planet's surface in order to be detected remotely.

An Earth-like planet is used to describe a planet similar in mass, radius, and temperature to Earth. The term Earth twin is usually reserved for an Earth-like planet with liquid water oceans and continental land masses.

Beyond the above generally accepted definitions are terms for exoplanets with no solar system counterparts. Super Earths are loosely defined as planets with masses between 1–10 $M_\oplus$, intended to include predominantly rocky planets. Mini Neptunes might be used for planets between about 10 and 15 Earth masses that have gas envelopes. The term water worlds has been used for planets that have 25% or more water by mass. Carbon planets refer to planets that contain more carbon than silicon, planets expected to form in an environment where elemental carbon is equal to or more abundant than oxygen. At present, observations cannot always distinguish among the above planet types.

Apart from planet types, there is not yet a definitive definition of atmosphere types (but see [8]). We do, however, expect atmospheres on different exoplanets to show a wide diversity, just as the orbits, masses, and radii of known exoplanets do. Planetary atmospheres originate either from direct capture of gas from the protoplanetary disk (for massive planets) or from outgassing during planetary accretion (for low-mass planets). The former mechanism is accepted for giant planets like Jupiter and Saturn and the latter is expected for terrestrial planets like Earth and Venus. The final atmospheric mass and composition will depend on the net of atmospheric sources versus. atmospheric sinks. The sources, both from direct capture and from outgassing, depend on the planet's location in the protoplanetary disk during formation, due to the compositional gradients in the disk. The atmospheric sinks include atmospheric escape and sequestering of gases in oceans. The range of atmospheric composition has yet to be uncovered theoretically and observationally.

The formalism presented in this book is intended to give you, the reader, knowledge to understand and interpret observations of any kind of exoplanet atmosphere.

REFERENCES

For further reading

For a review article on exoplanet atmospheres that summarizes observational and theoretical advances as of 2010:

- Seager, S., and Deming. D. 2010. "Exoplanet Atmospheres." Ann. Rev. Astron. Astrophys. 48, 631–672.

References for this chapter

1. Turnbull, M. C., Traub, W. A., Jucks, K. W., Woolf, N. J., Meyer, M. R., Gorlova, N., Skrutskie, M. F., and Wilson, J. C. 2006. "Spectrum of a Habitable World: Earthshine in the Near-Infrared." Astrophys. J. 644, 551–559.

2. Pearl, J. C., and Christensen, P. R. 1997. "Initial Data from the Mars Global Surveyor Thermal Emission Spectrometer Experiment: Observations of the Earth." J. Geophys. Res. 102, 10875–10880.

3. Kalas, P. et al. 2008. "Optical Images of a Planet 25 Light Years from Earth." Science 322, 1345–1348.

4. Marois, C., Macintosh, B., Barman, T., Zuckerman, B., Song, I., Patience, J., Lafrenier, D., and Doyon, R. 2008. "Direct Imaging of Multiple Planets Orbiting the Star HR 8799." Science 322, 1348–1352.

5. Knutson, H. A., et al. 2007. "A Map of the Day-Night Contrast of the Extrasolar Planet HD 189733b." Nature 447, 183–186.

6. http://exoplanet.eu/

7. Levine, M., Shaklan, S., and Kasting, J. 2006. "Terrestrial Planet Finder Science and Technology Definition Team (STDT) Report." NASA/JPL Document D-34923.

8. Seager, S., and Deming. D. 2010. "Exoplanet Atmospheres." Ann. Rev. Astron. Astrophys. 48, 631–672.

Chapter Two

Intensity and Flux

2.1 INTRODUCTION

The main goal of this book is to understand the origin of and physical processes affecting a planetary spectrum. We shall see just how much information can be derived from a planetary spectrum: the kinds of gases and solid particles present, the planetary albedo, and constraints on the vertical temperature structure. We begin with concepts and variables used to quantify radiation traveling through a planetary atmosphere.

As a foundational language for exoplanet atmospheres, these radiation terms are so important that we spend a chapter defining them carefully. This is especially important because the definitions of intensity, surface flux, and flux at Earth are used differently in other books and in the literature. In this chapter we will use the term "surface" to describe either the solid surface of a planet or the layers of the atmosphere from which radiation emerges.

2.2 INTENSITY

To begin with we need a description of radiation in the exoplanet atmosphere. We may think of radiation as energy in the form of photons traveling through the planet atmosphere. This radiation interacts with different matter particles. The conventional description of radiation considers the energy of a number of identical photons in a single beam of radiation, called the intensity I.

As the beam of radiation travels through the planet atmosphere, photons will be absorbed into and emitted out of the beam. Different parts of the planet, therefore, have different intensities, and the intensity varies with frequency. We cannot measure the intensity coming from a specific part of the interior or exterior of an exoplanet. This is because the exoplanet is so distant that the planet atmosphere cannot be spatially resolved. Nevertheless, I is a macroscopic parameter describing the sum of all microscopic processes going on in the beam of radiation. We therefore need to compute I in detail and carry it along through calculations of radiative transfer until we are ready to compute the final quantities of radiation that we are interested in.

The intensity I is formally defined as the amount of energy passing through a surface area dA, within a differential solid angle $d\Omega$ centered about $\hat{\mathbf{n}}$, per frequency

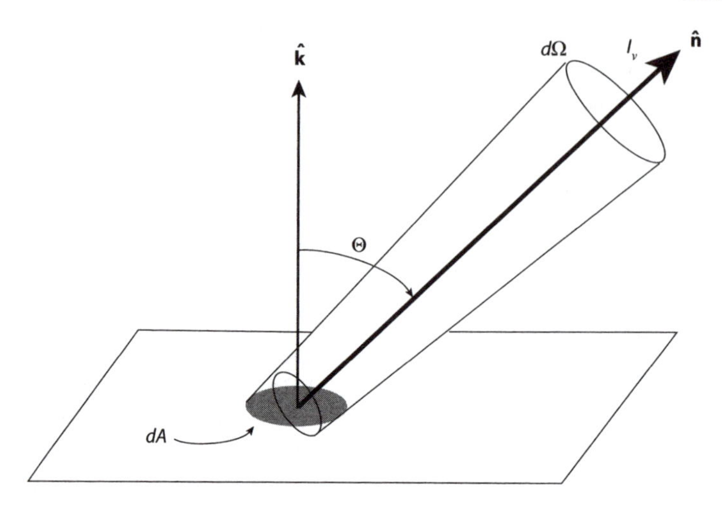

Figure 2.1 Definition of the specific intensity $I(\mathbf{x}, \hat{\mathbf{n}}, \nu, t)$.

interval, per unit time (Figure 2.1),

$$dE(\nu, t) = I(\mathbf{x}, \hat{\mathbf{n}}, \nu, t)\hat{\mathbf{n}} \cdot \hat{\mathbf{k}} d\Omega dA d\nu dt. \tag{2.1}$$

The SI units of I are J m^{-2} sr^{-1} s^{-1} Hz^{-1}. I is the intensity at location $\mathbf{x}$, going into direction $\hat{\mathbf{n}}$; directionality is implied despite the point that I is a scalar quantity.

We denote vectors in boldface, using italic for 1D and 2D vectors and roman for 3D vectors. When necessary to specify the direction of I or other scalars, we will denote a direction with a subscript.

The mean intensity is the intensity averaged over a solid angle,

$$J(\mathbf{x}, \nu, t) = \frac{1}{4\pi} \int_\Omega I(\mathbf{x}, \hat{\mathbf{n}}, \nu, t) d\Omega. \tag{2.2}$$

See Figure 2.2 for a definition of solid angle.

2.3 FLUX AND OTHER INTENSITY MOMENTS

The quantity of radiation that we do measure from exoplanets is related to the flux. The flux is the net flow of energy through an arbitrarily oriented surface area dA with normal $\hat{\mathbf{n}}$, per frequency interval, per unit time. The flux $\mathbf{F}$ is derived from the intensity in direction $\hat{\mathbf{n}}$ integrated over solid angle Ω ,

$$\mathbf{F}(\mathbf{x}, \nu, t) = \int_\Omega I(\mathbf{x}, \hat{\mathbf{n}}, \nu, t)\hat{\mathbf{n}} d\Omega. \tag{2.3}$$

$\mathbf{F}(\mathbf{x}, \nu, t)$ has units of J m^{-2} s^{-1} Hz^{-1}. $\mathbf{F}(\mathbf{x}, \nu, t)$ is defined at each location $\mathbf{x}$ in the planetary atmosphere.

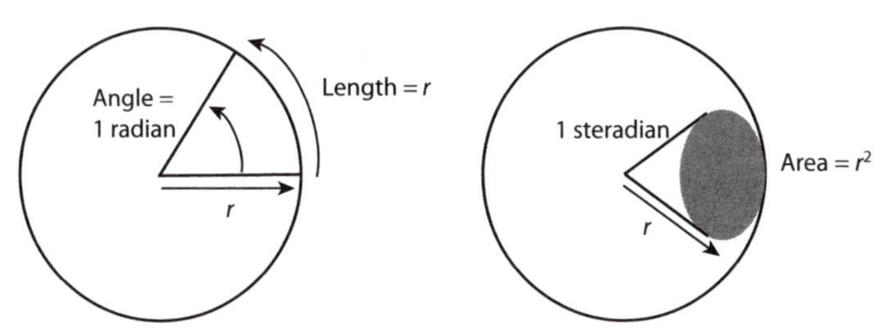

Figure 2.2 Definition of a radian (left panel) and a steradian (right panel). A steradian is related to the surface area of a sphere as a radian is related to the circumference of a circle. In 2D 1 radian is the angle subtended at the center of a circle by an arc length equal to the radius of a circle. In 3D 1 steradian is a measure of solid angle and is the solid angle subtended at the center of a sphere of radius r having an area r^2.

Flux is a vector and so we may write it in terms of its vector components. In a rectangular Cartesian coordinate system,

$$\mathbf{F}(\mathbf{x}, \nu, t) = F_i(\mathbf{x}, \nu, t)\hat{\mathbf{i}} + F_j(\mathbf{x}, \nu, t)\hat{\mathbf{j}} + F_k(\mathbf{x}, \nu, t)\hat{\mathbf{k}}. \tag{2.4}$$

In describing planetary atmospheres we are usually interested in the flux in one direction, the direction toward us, the observer. It is customary to take one component of the flux vector $F_k(\mathbf{x}, \nu, t)\hat{\mathbf{k}}$, writing only the magnitude $F_k(\mathbf{x}, \nu, t)$, a scalar quantity. We are essentially describing the energy flow in one direction

$$\boxed{F_k(\mathbf{x}, \nu, t) = \frac{dE(\nu, t)}{dA d\nu dt} = \int_\Omega I(\mathbf{x}, \hat{\mathbf{n}}, \nu, t)\hat{\mathbf{n}} \cdot \hat{\mathbf{k}} d\Omega.} \tag{2.5}$$

To further complicate the issue, the directional subscript is usually dropped so as to just write the flux in the direction of the observer as $F(\mathbf{x}, \nu, t)$.

The flux is also called the first moment of intensity. The zeroth moment of the intensity is the mean intensity, defined above in equation [2.2]. The second moment of intensity is

$$\boxed{\mathbf{K}(\mathbf{x}, \nu, t) = \frac{1}{4\pi} \int_\Omega I(\mathbf{x}, \hat{\mathbf{n}}, \nu, t)\hat{\mathbf{n}}\hat{\mathbf{n}} d\Omega.} \tag{2.6}$$

$\mathbf{K}(\mathbf{x}, \nu, t)$ is a tensor quantity related to the radiation pressure tensor $\mathbf{P}(\mathbf{x}, \nu, t)$, by $\mathbf{P} = \frac{1}{c}\mathbf{K}$, where c is the speed of light. When we come to use K in Chapter 6, we will take the magnitude of the tensor component of interest, a scalar quantity (as we have described for flux in equations [2.4] and [2.5]).

2.4 SURFACE FLUX

For exoplanet atmospheres we are most interested in the flux emerging from the planet surface. We call this kind of flux the surface flux. "Surface" refers to the layers of the planet atmosphere where most of the radiation originates without further interaction with gas, liquid, or solid particles. Surface flux is the outgoing

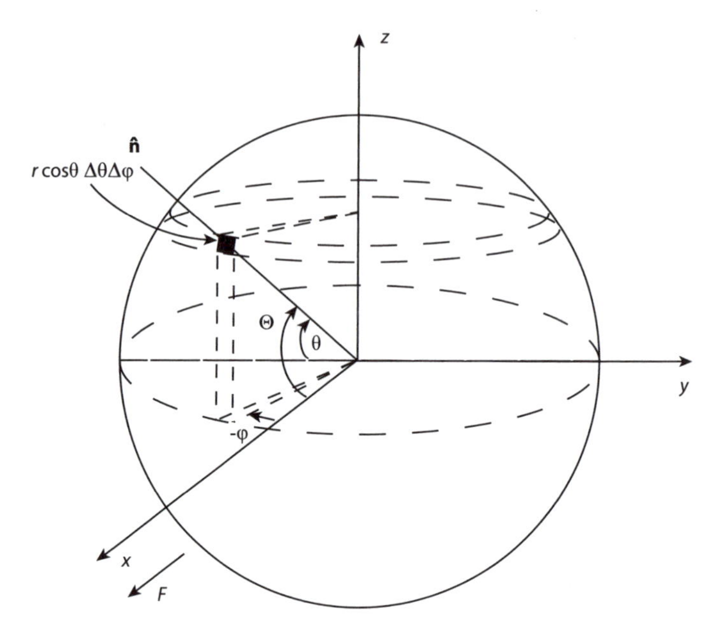

Figure 2.3 The spherical polar coordinate system using latitude θ and longitude ϕ. For flux computed along the x-axis, i.e., where the x-axis is $\hat{\mathbf{k}}$, and $\hat{\mathbf{n}} \cdot \hat{\mathbf{k}}$ is $\cos\Theta$.

flux at a particular point on the planet's surface. We will use $F_{\rm S}$ to distinguish the surface flux from the flux F which may be defined in a small volume anywhere in the planet atmosphere. Here we emphasize a subtle but important point. That is, as described in Section 2.3, we take only the magnitude of the flux vector in the outgoing direction of interest, a scalar quantity. So we will use the symbol $F_{\rm S}$ to denote surface flux in the outgoing direction of interest.

To describe the surface flux we first choose the spherical polar coordinate system (r, θ, ϕ) shown in Figure 2.3. In this coordinate system θ is the latitude and ϕ the longitude. This coordinate system is useful when considering a surface in a planetary latitude and longitude system. For example, $\theta = 0$ is defined according to the planet-star ecliptic plane, so that $(\theta = 0, \phi = 0)$ corresponds to the planetary substellar point, the point that receives the most stellar radiation. On Earth, the equator $(\theta = 0)$ is defined by the rotational axis of Earth.

We now go through the terms on the right-hand side of the flux definition equation [2.5] in order to derive an expression for the surface flux. In the spherical polar coordinate system we replace the vector $\mathbf{x}$ with (r, θ, ϕ), where again θ and ϕ refer to coordinates on the planetary surface. Furthermore, because we are only interested in the surface intensity, not the intensity at different altitudes in the planet atmosphere, we drop the reference to r, implicitly assuming $r = R_{\rm p}$. Now, the intensity $I(\mathbf{x}, \hat{\mathbf{n}}, \nu, t)$ has two vectors, one of which we have replaced by (θ, ϕ). For the second vector, the direction of $\hat{\mathbf{n}}$, we use two more angles θ_n and ϕ_n, with origin at $\hat{\mathbf{n}}$. We now write the surface intensity as $I_{\rm S}(\theta, \phi, \theta_n, \phi_n, \nu, t)$

We now move to the second term in equation [2.5] and note that $\hat{\mathbf{n}} \cdot \hat{\mathbf{k}} = \cos\Theta_n$.

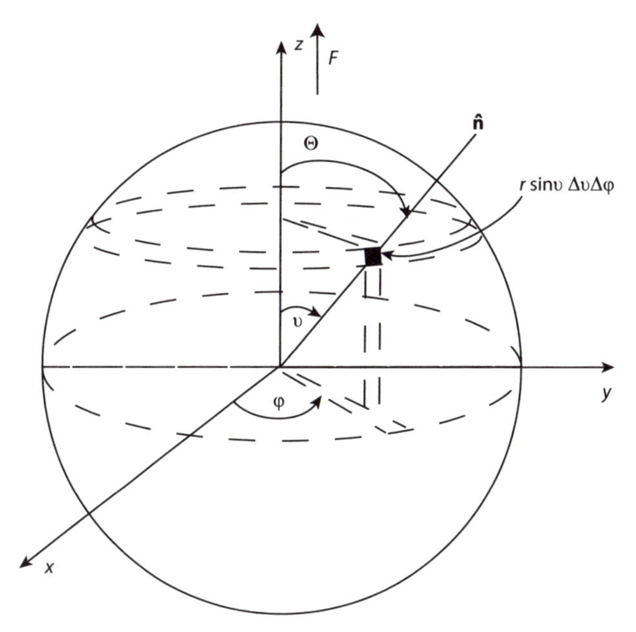

Figure 2.4 The spherical polar coordinate system using colatitude ϑ and longitude ϕ. For flux computed along the z-axis (the $\hat{\mathbf{k}}$ direction), $\hat{\mathbf{n}} \cdot \hat{\mathbf{k}} = \cos\vartheta$.

From Figure 2.3, and using the spherical cosine law, we have $\hat{\mathbf{n}} \cdot \hat{\mathbf{k}} = \cos\Theta_n = \cos\theta_n \cos\phi_n$. Again, use of θ_n and ϕ_n means we no longer need $\hat{\mathbf{n}}$ or $\hat{\mathbf{k}}$ in our description of intensity.

Lastly, the differential solid angle for surface flux at a given location on the planet surface is defined by

$$d\Omega_n = \cos\theta_n d\theta_n d\phi_n. \tag{2.7}$$

The surface flux is

$$\boxed{F_{\rm S}(\theta, \phi, \nu, t) = \int_{-\pi/2}^{\pi/2} \int_{-\pi/2}^{\pi/2} I_{\rm S}(\theta, \phi, \theta_n, \phi_n, \nu, t) \cos\phi_n \cos^2\theta_n d\theta_n d\phi_n.} \tag{2.8}$$

Recall that here $F_{\rm S}(\theta, \phi, \nu, t)$ is the scalar of the vector component of flux traveling in a given direction. Into which direction is the flux traveling? In this example, the surface flux is traveling out from the planet along direction $\hat{\mathbf{k}}$. We have further specified that the flux is at location θ, ϕ.

In some situations it is easier to solve for the surface flux (or surface intensity) in a coordinate system with one of the angles originating at the z-axis, as shown in Figure 2.4. The difference from the longitude-latitude spherical polar coordinate system is that the colatitude $\vartheta = 90° - \theta$ is used instead of the latitude θ as one of the independent variables. In this coordinate system, the substellar point is at the pole, and the flux is specified along the z-axis. The benefit of this system is that

$$\hat{\mathbf{n}} \cdot \hat{\mathbf{k}} = \cos\Theta_n = \cos\vartheta_n. \tag{2.9}$$

Furthermore, for problems symmetric in ϕ equations can be further simplified. The surface flux in this colatitude coordinate system is

$$\boxed{F_{\rm S}(\vartheta, \phi, \nu, t) = \int_0^{2\pi} \int_0^{\pi/2} I_{\rm S}(\vartheta, \phi, \vartheta_n, \phi_n, \nu, t) \cos \vartheta_n \sin \vartheta_n d\vartheta d\phi.} \quad (2.10)$$

We pause here to emphasize that the surface flux at (θ, ϕ) or (ϑ, ϕ) is usually written without reference to the surface coordinates, for example, by

$$F_{\rm S}(\nu, t) = \int_{-\pi/2}^{\pi/2} \int_{-\pi/2}^{\pi/2} I_{\rm S}(\theta, \phi, \nu, t) \cos \phi \cos^2 \theta d\theta d\phi. \quad (2.11)$$

or

$$F_{\rm S}(\nu, t) = \int_0^{2\pi} \int_0^{\pi/2} I_{\rm S}(\vartheta, \phi, \nu, t) \cos \vartheta \sin \vartheta d\vartheta d\phi. \quad (2.12)$$

Here θ and ϕ or ϑ and ϕ refer to the solid angle integration of the intensity, and even though the surface flux is for a specific surface element it is not specified. Although less precise, for a planet with uniform surface intensity this surface flux description is adequate, and we will be guilty of adopting it.

As an example of surface flux, we will compute the surface flux for a planet with uniform intensity $I_{\rm S}(\theta, \phi, \nu, t) = I_{\rm S}(\nu, t)$. This uniform intensity may be pulled out of the integrand in either equation [2.8] or equation [2.10] and the integrals performed to yield

$$\boxed{F_{\rm S}(\nu, t) = \pi I_{\rm S}(\nu, t).} \quad (2.13)$$

Sometimes it is useful to integrate the flux over all wavelengths, and we denote this wavelength-independent flux without a ν-dependence

$$\boxed{F_{\rm S}(t) = \textstyle\int_0^\infty F_{\rm S}(\nu, t) d\nu.} \quad (2.14)$$

2.5 OBSERVED FLUX

What form of radiation are we able to measure at Earth? At Earth we see the planet as a point source, that is, as spatially unresolved. All radiation from the exoplanet hemisphere is averaged into a single value of flux. Recall that the surface intensity depends on location on the planet surface described by θ and ϕ. We must integrate the surface intensity from each location on the planet into a quantity we can actually measure.

At Earth we are measuring the energy collected from a planet subtended by an angle Ω, by a detector of a given area, in a frequency interval, and during some interval of time. We will denote the flux measured at the detector by $\mathcal{F}_\oplus(\nu, t)$. To derive $\mathcal{F}_\oplus(\nu, t)$ for a distant planet, we must realize that the integration over solid angle is at the detector and not at the planet surface.

In order to derive the surface flux at Earth we return to the definition of flux, equation [2.5], and use Figure 2.5. In the derivation of the flux at Earth it is convenient to use the colatitude spherical polar coordinate system (Figure 2.4). In this

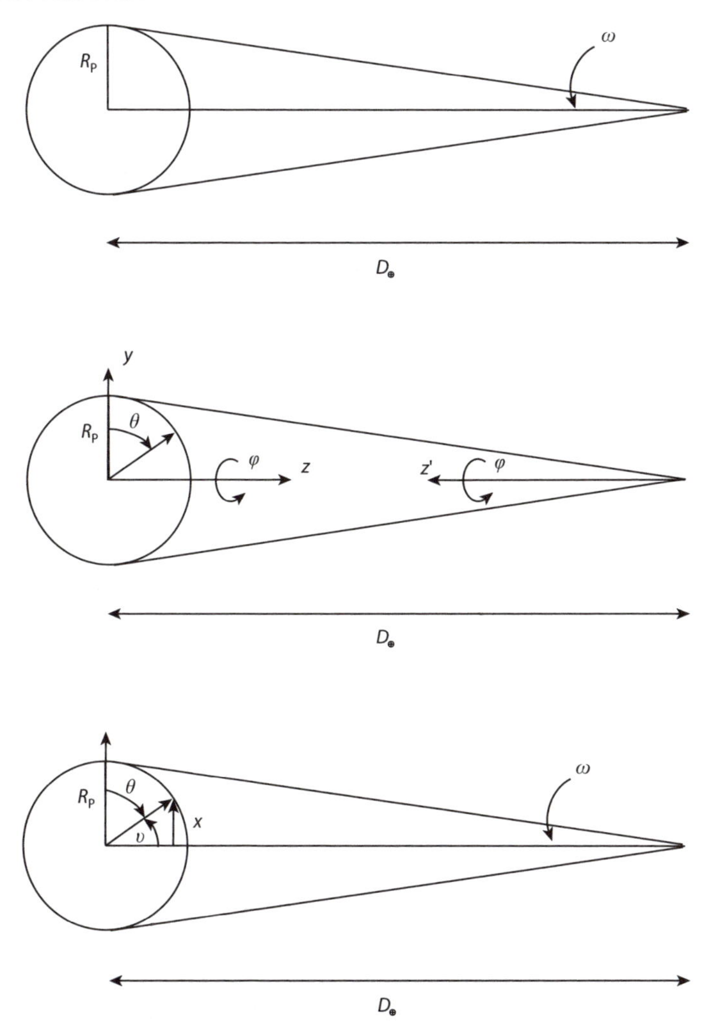

Figure 2.5 Derivation of the planet flux observed at Earth, $\mathcal{F}_{\oplus}(\nu, t)$.

coordinate system, the flux is defined as leaving the planet along the z-axis, and the z-axis is pointing toward the detector at Earth. Recall that the angle ϕ is defined about the z-axis. From the detector at Earth, the planet subtends the solid angle defined by

$$\Omega = \int_0^{2\pi} \int_0^{R_{\rm p}/D_{\oplus}} \sin \omega d\omega d\phi. \tag{2.15}$$

A critical point in the derivation of measured flux is to recognize that the ϕ component of angle subtended by the planet at Earth is equivalent to the angle ϕ in the colatitude coordinate system on the planet (see the middle panel of Figure 2.5).

In the coordinate system of the detector at Earth we have for the flux

$$\mathcal{F}_{\oplus}(\nu,t) = \int_0^{2\pi}\int_0^{R_{\rm p}/D_{\oplus}} I_{\rm S}(\vartheta,\phi,\nu,t)\cos\omega\sin\omega d\omega d\phi. \tag{2.16}$$

Note that here we have retained the description of intensity in the coordinate system of the planet rather than change to the coordinate system at the detector.

We now convert the term $\cos\omega\sin\omega d\omega$ in equation [2.16] into the coordinate system of the planet. Using Figure 2.5, we see that

$$\omega = \frac{R_{\rm p}}{D_{\oplus}}\sin\vartheta. \tag{2.17}$$

Additionally, $d\omega = (R_{\rm p}/D_{\oplus})\cos\vartheta d\vartheta$. In the limit that $R_{\rm p} \ll D_{\oplus}$, we have $\omega \ll 1$ and can make some approximations. Using a Taylor expansion, $\sin\omega \sim \omega$ and $\cos\omega \sim 1-\omega^2/2 \sim 1$. In this limit of $R_{\rm p} \ll D_{\oplus}$ we can also approximate $D_{\oplus}$ as the distance from the surface of the star to the observer at Earth. Equation [2.16] then becomes

$$\boxed{\mathcal{F}_{\oplus}(\nu,t) = \left(\frac{R_{\rm p}}{D_{\oplus}}\right)^2 \int_0^{2\pi}\int_0^{\pi/2} I_{\rm S}(\vartheta,\phi,\nu,t)\cos\vartheta\sin\vartheta d\vartheta d\phi.} \tag{2.18}$$

We emphasize that $\mathcal{F}_{\oplus}(\nu,t)$ has the same dimensions as the surface flux in equation [2.10] and that in our limit that the planet is very distant from Earth the two fluxes are related by (see equation [2.12])

$$\boxed{\mathcal{F}_{\oplus}(\nu,t) = \left(\frac{R_{\rm p}}{D_{\oplus}}\right)^2 F_{\rm S}(\nu,t).} \tag{2.19}$$

2.6 LUMINOSITY AND OUTGOING ENERGY

Luminosity $L(t)$ is defined as the rate at which a planet radiates energy in all directions. Another way to think about luminosity is as the summation of flux passing through a closed surface encompassing the planet. For uniform flux $F_{\rm S}(\theta,\phi,t) = F_{\rm S}(t)$ and for a planet that radiates equally in all directions,

$$L(t) = \int_{\rm A} F_{\rm S}(t) dA, \tag{2.20}$$

in units of J s^{-1}. Using the surface element of a sphere

$$dA = R_{\rm p}^2 \sin\theta d\theta d\phi, \tag{2.21}$$

we have

$$L(t) = R_{\rm p}^2 \int_0^{2\pi}\int_0^{\pi} F_{\rm S}(t)\sin\theta d\theta d\phi, \tag{2.22}$$

and integrate to find

$$\boxed{L(t) = 4\pi R_{\rm p}^2 F_{\rm S}(t).} \tag{2.23}$$

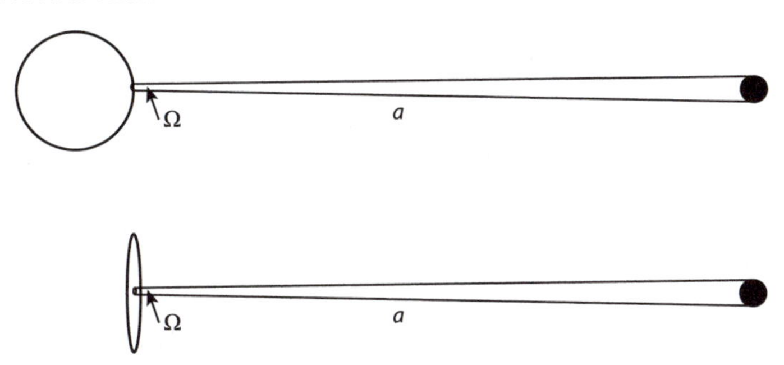

Figure 2.6 Incident flux on a sphere and a disk. The star subtends an angle Ω as seen from each location on the planet. To find the total energy per unit time incident on the planet, the flux at each location must be summed up over the planet hemisphere. Here a is semimajor axis.

For giant exoplanets $L(t)$ is useful to describe the flux coming from the planet interior that is incident on the lower boundary of the atmosphere. This interior flux is considered to be uniform around the planet at the bottom of the atmosphere, and originates from gravitational potential energy left over from the planet's nascent contraction. $L(t)$ may also be used for the rate at which energy leaves the planet in all directions. Often, however, we want to compute the total flux passing through only one hemisphere of the planet, not the entire planet. This is because, when observing any exoplanet, flux from only one hemisphere is visible to us at any given time. For a planet that does not radiate uniformly in all directions from all locations, we have to take care not to use the luminosity $L(t)$, but instead to use energy per unit time,

$$E_{\mathrm{S}}(t) = R_{\mathrm{p}}^2 \int_0^{\pi} \int_0^{\pi/2} F_{\mathrm{S}}(\theta, \phi, t) \sin\theta d\theta d\phi, \tag{2.24}$$

and, assuming uniform flux, integrate to find

$$E_{\mathrm{S}}(t) = 2\pi R_{\mathrm{p}}^2 F_{\mathrm{S}}(t). \tag{2.25}$$

We now turn from energy per unit time leaving the planet to energy per unit time incident on the planet.

2.7 INCIDENT FLUX AND INCIDENT ENERGY

For exoplanets, the amount of radiation from the star reaching the planet is critical. The radiation from the star heats the planet and ultimately governs the global energy balance. The stellar heating also drives mass motion in the planet atmosphere. We therefore want to know the amount of flux or energy per unit time from the parent star that falls on the planet surface. We call this incident radiation, incident intensity, incident flux, or incident energy, depending on the context. More generally, we may sometimes even call the incident radiation "irradiation." To derive an expression for incident flux we want to consider the solid angle subtended by the star on a surface element of the planet, as shown in Figure 2.6. We want to compute the flux from the star at a given location on the planet.

We use equation [2.18], an equation we previously derived for the planet's flux observed at Earth. We now want to know the flux of the star as observed from a location on the planet. In equation [2.18], we therefore replace $R_{\rm p}$ with R_* and $D_\oplus$ with the planet semimajor axis a. We also replace the planet surface intensity $I_{\rm S}$ with the stellar surface intensity $I_{{\rm S},*}$,

$$\mathcal{F}_{\rm inc}(\vartheta, \phi, \nu, t) = \left(\frac{R_*}{a}\right)^2 \int_0^{2\pi} \int_0^{\pi/2} I_{{\rm S},*}(\nu, t)(\vartheta, \phi, \nu, t) \cos\vartheta \sin\vartheta d\vartheta d\phi. \tag{2.26}$$

To simplify this equation, we can fairly assume that the stellar intensity is uniform across the star's surface, yielding

$$\mathcal{F}_{\rm inc}(\nu, t) = \left(\frac{R_*}{a}\right)^2 \pi I_{{\rm S},*}(\nu, t), \tag{2.27}$$

where we recall that $\mathcal{F}(\nu, t) = \pi I(\nu, t)$ from equation [2.13], and then also have

$$\mathcal{F}_{\rm inc}(\nu, t) = \left(\frac{R_*}{a}\right)^2 F_{{\rm S},*}(\nu, t). \tag{2.28}$$

We have discussed the incident flux at one location of the planet (equations [2.27] and [2.28]). For many applications we will want to know the *total* incident energy per unit time, and we must integrate over the surface of the planet. We cannot simply multiply by $2\pi R_{\rm p}^2$, the surface area of the planet hemisphere. This is due to the fact that only the substellar point on the planet receives the full amount of stellar flux. The planet locations away from the substellar point receive an amount reduced by $\hat{\mathbf{n}} \cdot \hat{\mathbf{k}} = \cos\Theta$ (see Figure 2.3). We are familiar with this concept on Earth because the poles on Earth receive much less sunlight than the equatorial region. We proceed to integrate equation [2.28] over the surface area of one hemisphere

$$E_{\rm inc}(\nu, t) = \left(\frac{R_*}{a}\right)^2 F_{{\rm S},*}(\nu, t) \int_0^{2\pi} \int_0^{\pi/2} R_{\rm p}^2 \cos\vartheta \sin\vartheta d\vartheta d\phi, \tag{2.29}$$

to find

$$E_{\rm inc}(\nu, t) = \pi R_{\rm p}^2 \left(\frac{R_*}{a}\right)^2 F_{{\rm S},*}(\nu, t). \tag{2.30}$$

We also define a total incident energy over all wavelengths by

$$\boxed{E_{\rm inc}(t) = \pi R_{\rm p}^2 \left(\frac{R_*}{a}\right)^2 F_{{\rm S},*}(t).} \tag{2.31}$$

We have just introduced the total flux and total energy incident on an exoplanet. For plane-parallel radiation from a distant point source, we will find use for a description of the incident stellar intensity as

$$I_*(\vartheta, \phi, \nu, t) = I_0 \delta(\vartheta - \vartheta_0)\delta(\phi - \phi_0). \tag{2.32}$$

Here, the star is in the direction θ_0, ϕ_0 from the surface normal. We consider that the star is far enough away from the planet so that the only the rays in one direction are incident on the planet. At each location on the planet, the star is in a different position on the sky, and hence θ_0 and ϕ_0 are different for each surface element. The incident flux at a given location (θ, ϕ) on the planet surface is

$$\begin{aligned}\mathcal{F}_{\rm inc}(\vartheta, \phi, \nu, t) &= \left(\frac{R_*}{a}\right)^2 \int_0^{2\pi} \int_0^{\pi/2} I_0 \delta(\vartheta - \vartheta_0)\delta(\phi - \phi_0) \cos\vartheta \sin\vartheta d\vartheta d\phi \\ &= \left(\frac{R_*}{a}\right)^2 \cos\vartheta_0 \sin\vartheta_0 I_0. \end{aligned} \quad (2.33)$$

2.8 BLACK BODY INTENSITY AND BLACK BODY FLUX

Black body flux can be used to estimate both the flux incoming to and outgoing from an exoplanet. We will assume that the reader is already familiar with a black body radiator and the derivation of the intensity black body radiation. Here we aim to describe the black body intensity and flux in our framework and its relevance to exoplanetary atmospheres. A black body is a "perfect" radiator that absorbs all radiation incident on it and reemits radiation in a frequency spectrum depending only on its temperature T. Black body radiation is furthermore isotropic and hence has no $\hat{\mathbf{n}}$-dependence. The black body radiation depends only on temperature and frequency and can be described by $B(T, \nu)$.

The Planck function describes the intensity of black body radiation,

$$B(T, \nu) = \frac{2h\nu^3}{c^2} \frac{1}{e^{h\nu/kT} - 1}. \quad (2.34)$$

Here h is Planck's constant, k is Boltzmann's constant, and c is the speed of light. The black body intensity by definition has the same dimensions as the intensity I, units of J m^{-2} sr^{-1} s^{-1} Hz^{-1}. Because the temperature varies with location in the planet atmosphere, and possibly with time, we can also write the black body intensity as

$$B(\mathbf{x}, \nu) = \frac{2h\nu^3}{c^2} \frac{1}{e^{h\nu/kT} - 1}. \quad (2.35)$$

We emphasize that our description of black body radiation is radiation per frequency bin $(d\nu)$. Black body radiation per wavelength bin $d\lambda$ must include the conversion factor $d\nu = -c/\lambda^2 d\lambda$, where the $-$ sign can be absorbed into $d\lambda$,

$$B(T, \lambda) = \frac{2hc^2}{\lambda^5} \frac{1}{e^{hc/\lambda kT} - 1}. \quad (2.36)$$

The black body flux can be computed from the black body intensity B using the flux definition in equation [2.5]. Black body radiation is isotropic, that is, uniform in all directions, so that outward from one hemisphere

$$F_{\rm S}(T, \nu) = \pi B(T, \nu). \quad (2.37)$$

We will always use B as black body intensity and πB as black body flux.

Stellar and planet atmospheres can be approximated as black body radiators, even though a black body is a highly idealized construct. Figure 2.7 shows the black

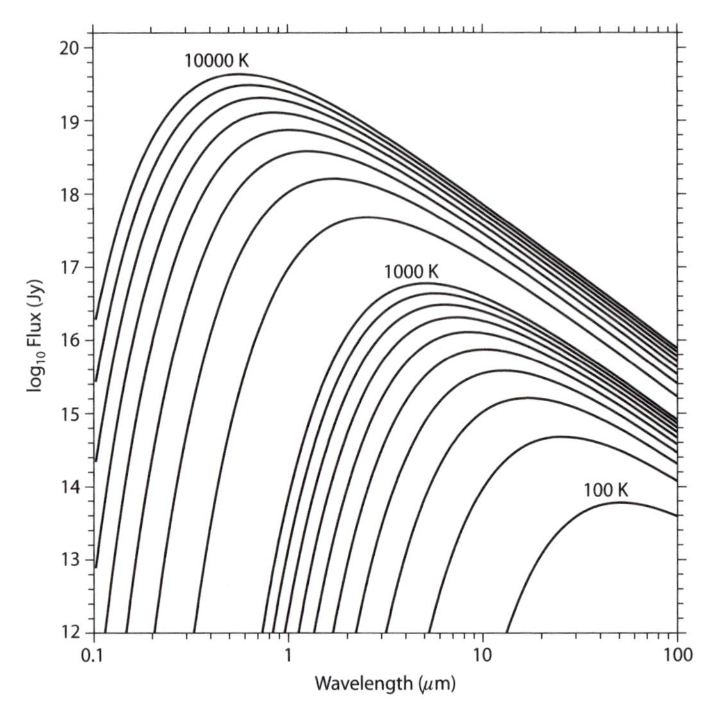

Figure 2.7 Black body surface fluxes (in units of Jy = 10^{-26} W m^{-2} Hz^{-1}) for a range of temperatures. The black body fluxes are spaced by 100 K in the range 100 to 1000 K and by 1000 K in the range 1000 to 10,000 K. The Sun may be represented by a 5750 K black body, while Earth can be approximated by a 300 K black body.

body surface flux for different temperatures. The effective temperature of stars with known planets ranges from 3000 to 6000 K, and known planets have effective temperatures ranging from 60 to over 2000 K. The magnitude and frequency peak of the black body flux increase with decreasing temperature.

2.9 LAMBERT SURFACE

A Lambert surface is often used to approximate the reflectivity of planetary bodies. Understanding a Lambert surface helps to understand the difference between two fundamental concepts: the intensity emanating from an object and the intensity measured by a distant observer. A Lambert surface is a surface that scatters intensity isotropically (i.e., equally in all directions).

What does it mean conceptually for a surface to scatter equal intensity in all directions? Equal intensity means that the apparent surface brightness of an area element is the same from any viewing angle. As an example, consider a sheet of white paper illuminated from above. A piece of white paper is an approximate plane Lambertian reflector. In other words, if you hold a sheet of paper up to

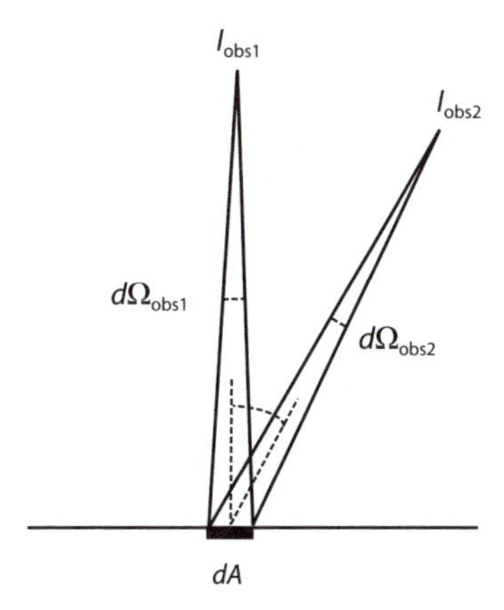

Figure 2.8 Intensity observed by a distant viewer. The angle subtended by dA is smaller at observer 2 than at observer 1.

the light and view it from different directions, the paper appears to be the same brightness. A second example is holding a lightbulb up to a white painted wall. The wall is brightest perpendicular to the light bulb, but the bright spot's location and extent does not change with changing viewing angle.

We will now expand on the conceptual description of a Lambert surface, by showing that a plane Lambert surface, illuminated from one direction perpendicular to the surface normal, has the same apparent brightness as viewed from any direction. Figure 2.8 shows the intensity viewed by an observer looking perpendicular to the plane and an observer viewing at an angle ϑ from the surface normal. Both observers are viewing the same differential area dA on the Lambert surface. We note that the distance to the observer is much greater than the size of the surface element. From the definition of intensity (equation [2.1]) we have the intensity measured by observer 1,

$$I_{\text{obs1}} = \frac{dE_{\text{obs1}}}{dA d\Omega_{\text{obs1}} \hat{\mathbf{n}}_{\text{obs1}} \cdot \hat{\mathbf{k}}}, \tag{2.38}$$

and, similarly, the intensity measured by observer 2 is

$$I_{\text{obs2}} = \frac{dE_{\text{obs2}}}{dA d\Omega_{\text{obs2}} \hat{\mathbf{n}}_{\text{obs2}} \cdot \hat{\mathbf{k}}}. \tag{2.39}$$

We have by definition of the coordinate system (Figure 2.8) that $\hat{\mathbf{n}}_{\text{obs1}} \cdot \hat{\mathbf{k}} = 1$ and $\hat{\mathbf{n}}_{\text{obs2}} \cdot \hat{\mathbf{k}} = \cos\vartheta$. In other words, the solid angle subtended by the surface element dA at observer 2 is smaller than the solid angle subtended by dA at observer 1. This is more easily apparent for the extreme case of $\vartheta \sim 90^\circ$ where the solid angle subtended approaches zero. We also have the relationship

$$dE_{\text{obs2}} = \cos\vartheta dE_{\text{obs1}}, \tag{2.40}$$

because for isotropic scattering the energy (or number of photons) drops off as $\cos\vartheta$ away from the normal (i.e., away from the direction of incident radiation).

Putting the above equations together, we have, from observer 2

$$I_{\rm obs2} = \frac{dE_{\rm obs1}\cos\vartheta}{dAd\Omega_{\rm obs2}\cos\vartheta}. \tag{2.41}$$

From comparison with equation [2.38] we see that

$$I_{\rm obs1} = I_{\rm obs2}. \tag{2.42}$$

For a plane Lambert surface the intensity—or brightness—appears the same from all directions: the smaller number of photons emerging by the slant direction is compensated by the smaller subtended angle of an area element.

2.10 SUMMARY

We have presented fundamental definitions and concepts needed to describe radiation traveling through a planetary atmosphere. We began with a precise definition of the intensity and flux to be used throughout this book, quantities that have a variety of definitions in the literature (see exercise 1). We made a careful investigation of the surface flux on a planet as compared to the observed flux at Earth, and showed that these are the same if the planet is far enough away and if the planet's intensity is uniform across the planet's surface. Here the word "surface" might refer to a solid planetary surface like Earth's, or, in the case of a giant planet, it might refer to the deep atmosphere layers that become optically thick (akin to the photosphere of the Sun). We continued to define the quantities of outgoing luminosity and incident stellar flux, as well as the challenging concept of the Lambertian surface. With a handle on the fundamental definitions and concepts we are ready to embark on the task of understanding the basic physical characteristics and observables of exoplanets, planetary temperatures, albedos, and flux ratios.

REFERENCES

For further reading

For further introduction to the definitions of intensity, flux, and black body radiation:

- Chapter 1 in Rybicki, G. B., and Lightman, A. P. 1986. *Radiative Processes in Astrophysics* (New York: J. Wiley and Sons).
- Mihalas D. 1978. *Stellar Atmospheres* (2nd ed.; San Francisco: W. H. Freeman).
- Shu, F. 1991. *The Physics of Astrophysics. Vol. I, Radiation* (Mill Valley: University Science Books).

Reference for this chapter

1. McCluney, W. R. 1994. *Introduction to Radiometry and Photometry* (Norwood: Artech House).

Table 2.1 SI radiometry units.

Quantity	SI unit (abbr.)	Notes
Radiant energy	J	Energy
Radiant flux	W	Also called radiant power
Radiant intensity	W sr^{-1}	Power per unit solid angle
Radiance	W sr^{-1} m^{-2}	Power per unit solid angle per unit area
Irradiance	W m^{-2}	Power incident on a surface
Radiant exitance or Radiant emitance	W m^{-2}	Power emitted from a surface
Radiosity	W m^{-2}	Emitted + reflected power from a surface
Spectral radiance	W sr^{-1} m^{-3} or W sr^{-1} m^{-2} Hz^{-1}	
Spectral irradiance	W m^{-3} or W m^{-2} Hz^{-1}	

Table adapted from [1].

EXERCISES

1. List the variables introduced in this chapter that describe radiation, including their dimensions. The SI system has standard definitions for radiation terms that differ from the ones conventionally used for exoplanets. Compare the radiation terms used in this chapter to the SI radiation terms in Table 2.1

2. Explain the meaning of isotropic radiation. Show that for isotropic radiation the flux integrated over a hemispheric solid angle is $F = \pi I$ but that flux integrated over a solid angle is $F = 0$. Show that for isotropic radiation $I = J$.

3. Show that the intensity I does not depend on distance in a medium with no extinction or emission. Show that the flux $\mathcal{F}$ follows the inverse square law $\mathcal{F} \sim 1/d^2$, where d is distance away from the source, and that this distance dependency is not in conflict with the constancy of I.

4. Show that the general expression for solid angle (in units of steradians) is

$$\Omega = \frac{1}{R^2} \int_0^{\theta} 2\pi R \sin\theta R d\theta = 2\pi(1 - \cos\theta). \tag{2.43}$$

   What is the large-angle limit as $\theta \rightarrow \pi$? What is the small-angle limit as $\theta \rightarrow 0$?

5. Show that $\cos\Theta = \cos\phi\cos\theta$ in the spherical polar coordinate system in Figure 2.3.

6. Frequency or wavelength of maximum flux and energy from a black body.
   a. Derive Wien's Law as a function of frequency. Wien's Law describes the relationship between the peak of emission of a black body radiator and frequency. Wien derived his law from thermodynamic arguments but you may use the derivative of the black body radiation (equation [2.34]).

   b. Repeat part a, beginning with equation [2.36], to derive Wien's Law as a function of wavelength.

   c. Use Wien's Law to estimate the wavelength and frequency at which the Sun's emitted energy peaks. We will assume that the Sun's emitted flux can be approximated by a black body of temperature 5750 K. Are the frequency and wavelength the same or different? Explain.

   d. Repeat part b for an M star with a temperature of 3500 K.

7. For incident radiation from a star onto a planet, the expression for flux is different for an extended source compared to a point source (Section 2.7.) At what star-planet separation can the star reasonably be approximated as a point source?

Chapter Three

Temperature, Albedos, and Flux Ratios

3.1 INTRODUCTION

When observing a distant exoplanet, the only quantity we can measure is the radiation coming from the planet, in a form we have called flux. The temperature of exoplanet atmospheres, in contrast, is the quantity that is actually relevant for physics and chemistry of the exoplanet atmosphere. While only a single flux may be measured, real planets rarely have a single temperature throughout. Which temperature should represent the planet? Which temperature should represent the flux? In this chapter we explore the relationships among the planet flux, temperature, and albedo. We also derive planet-to-star flux ratios which are key for assessing what kind and size of telescope and instrumentation are required to detect an exoplanet—and whether or not an exoplanet is detectable by any means.

3.2 ENERGY BALANCE

The planet atmosphere temperature and albedo are related by the fundamental principle of conservation of energy. For planetary atmospheres the conservation of energy is described as "energy balance." The underlying concept is that no energy is created or destroyed in a planetary atmosphere. All of the energy in the planet atmosphere comes either from the parent star—in the form of absorbed incident radiation—or from the planetary interior.

We can describe the energy balance by

$$E_{\mathrm{out}}(t) = (1 - A_{\mathrm{B}})E_{\mathrm{inc}}(t) + E_{\mathrm{int}}(t). \tag{3.1}$$

Here $E_{\mathrm{out}}(t)$ is the energy per unit time leaving the planet, while $E_{\mathrm{inc}}(t)$ is the stellar energy per unit time incident on the planet and $E_{\mathrm{int}}(t)$ is the energy per unit time transferred to the atmosphere from the planetary interior. The factor $(1 - A_{\mathrm{B}})$ is the fraction of incident stellar energy absorbed to heat the atmosphere or surface (and subsequently reemitted as radiation at longer wavelengths). The fraction of incident stellar energy scattered back into space is the Bond albedo A_{B}.

The planet exists in an equilibrium with the incident stellar radiation where the heating with time is constant. (Planets in eccentric orbits will not have constant heating with time.) For the rest of this chapter we adopt this energy balance by dropping the time dependence of the planet flux and related quantities. We note that energy per unit time is technically called power.

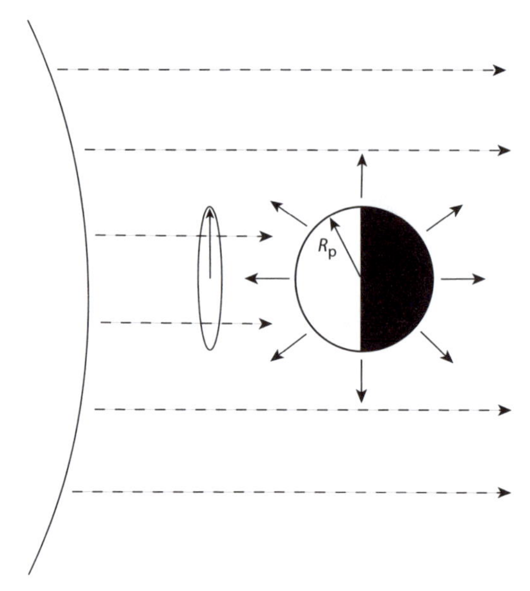

Figure 3.1 Schematic illustration for planetary energy balance. Stellar radiation falls onto the planet in an amount related to the planet's cross-sectional area of πR_p^2. A fraction $(1-A_B)$ of this radiation is absorbed by the planet. In this illustration the absorbed stellar radiation is advected around the planet and uniformly reemitted into 4π.

We can describe the energy balance in equation [3.1] using the planetary luminosity $L_p = E_{out}$ for outgoing energy and describing the incident stellar energy in terms of stellar flux and planet semimajor axis (see equations [2.23] and [2.29]),

$$4\pi R_p^2 F_{S,p} = (1 - A_B) F_{S,*} \left(\frac{R_*}{a}\right)^2 \pi R_p^2 + L_{p,int}. \tag{3.2}$$

On the right-hand side of equation [3.2] we see the two terms that describe the energy sources for the planetary atmosphere: the absorbed (and reradiated) stellar energy and the planet's own interior energy or luminosity $L_{p,int}$. The first term on the right-hand side is the energy absorbed by the planet (see Figure 3.1). It consists of the incident energy on the planet from the star, derived in equation [2.29], multiplied by the fraction of energy absorbed. This factor is $(1 - A_B)$, where A_B is the Bond albedo. We assume that this absorbed energy is eventually reradiated to space as long-wavelength radiation. Regarding the planet's interior luminosity $L_{p,int}$ (i.e., the interior energy), we recall the definition of luminosity: the flux passing through a surface encompassing the planet. The planet's internal luminosity has several different sources.

For giant planets such as Jupiter which are composed predominantly of hydrogen and helium, $L_{p,int}$ is the gradual loss of residual gravitational potential energy from the planet's formation. Indeed, Jupiter has an internal luminosity over twice as high as its luminosity from reradiated absorbed stellar energy. This is an indication of how long it takes for energy to travel from a giant planet interior out to space.

Earth's $L_{\rm p,int}$ arises partly from residual gravitational potential energy but mostly from decay of radioactive isotopes (of uranium, thorium, and potassium). For many planets, including the hot Jupiter exoplanets in very short-period orbits (less than four days) and terrestrial planets with typically limited amounts of interior energy, the luminosity from external heating overwhelms the interior luminosity by many orders of magntiude. We therefore write the energy balance equation

$$\boxed{4\pi R_{\rm p}^2 F_{\rm S,p} = (1 - A_{\rm B}) F_{\rm S,*} \left(\frac{R_*}{a}\right)^2 \pi R_{\rm p}^2.} \tag{3.3}$$

3.3 PLANETARY TEMPERATURES

There is no single temperature to describe the planet atmosphere. The temperature of a planet varies with altitude, with horizontal location around the planet, and possibly in time from day to night or from season to season. Which temperature should we use to describe the planet? Temperature is an important planet parameter since it governs to first order the chemical equilibrium state of the planet and hence the emergent spectrum. Here we describe three commonly used temperatures: the effective temperature, the equilibrium temperature, and the brightness temperature, and how they relate to the planet flux.

3.3.1 The Effective Temperature $T_{\rm eff}$

The effective temperature $T_{\rm eff}$ is used as a proxy for the global temperature of a planet atmosphere. $T_{\rm eff}$ is defined as the temperature of a black body of the same shape and at the same distance as the planet and with the same total flux as the planet. (Recall that here total flux refers to the flux integrated over all wavelengths or frequencies.) $T_{\rm eff}$ is in principle a measured quantity, taking the total flux of a planet, converting it to surface flux (i.e., flux radiated at the planet), and finding the temperature of a black body with the same total flux. Based on this definition, we can derive an equation for $T_{\rm eff}$ by: using the definition of black body flux (equation [2.37]); integrating the black body flux over all frequencies; and assuming the surface flux to be uniform across the object's surface

$$F_{\rm S} \equiv \int_0^\infty F_{\rm S}(\nu) d\nu = \pi \int_0^\infty B(T,\nu) d\nu \equiv \sigma_{\rm R} T_{\rm eff}^4. \tag{3.4}$$

Here $\sigma_{\rm R}$ is known as the Stefan-Boltzmann constant,

$$\sigma_{\rm R} = \frac{2\pi^5}{15} \frac{h}{c^2} \left(\frac{k}{h}\right)^4, \tag{3.5}$$

which has a value 5.670×10^{-8} J K^{-4} m^{-2} s^{-1}. $F_{\rm S}$ is the planetary surface flux (i.e., radiated at the planet surface) in units of J m^{-2} s^{-1}, as derived in Chapter 2. This relation is the Stefan-Boltzmann Law:

$$\boxed{F_{\rm S} = \sigma_{\rm R} T_{\rm eff}^4.} \tag{3.6}$$

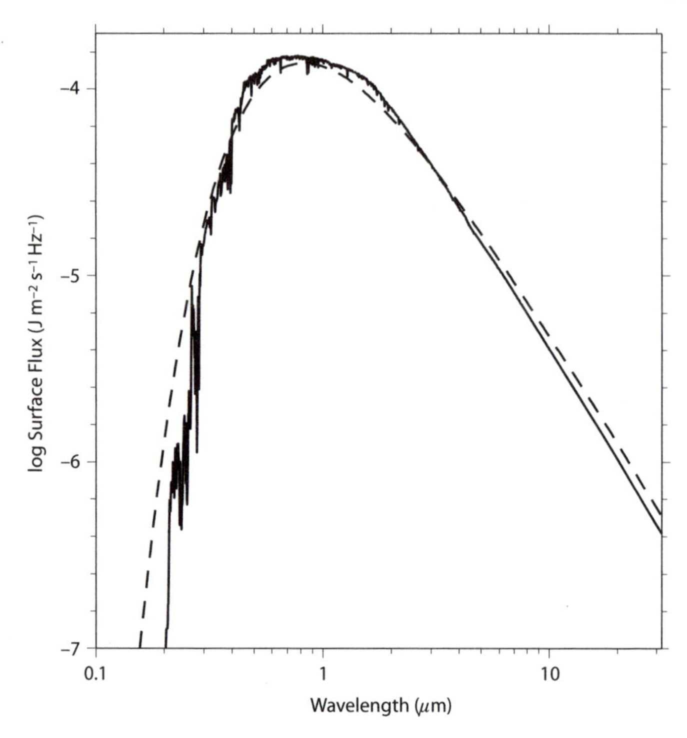

Figure 3.2 Illustration of the effective temperature definition. The theoretical surface flux of a $T_{\rm eff} = 6130$ K G0IV star is shown [1]. A black body with the same 6130 K effective temperature is shown by the dashed curve. The total stellar flux (integrated over all wavelengths) is equivalent to the total black body flux.

Figures 3.2 and 3.3 show the $T_{\rm eff}$ definition applied to two different stars. The first HD 149026 has a stellar flux similar to a black body, but the second, GJ 436, has a flux that departs significantly from a black body flux.

The effective temperature $T_{\rm eff}$ describes the global planet temperature at the altitude where the bulk of the radiation leaves the planet or star. This altitude is sometimes called the planet photosphere or surface, whether or not the altitude is at the planet surface. $T_{\rm eff}$ may be quite different from the surface temperature, as is the case for Venus. Venus's $T_{\rm eff}$ is $\sim$230 K, a value that comes from the cloud tops which obscure the planet's solid surface. Due to a strong greenhouse effect, Venus's surface temperature, at 730 K, is almost 500 K hotter than its $T_{\rm eff}$.

$T_{\rm eff}$ varies widely for planets. The coldest planets in our own solar system, Uranus and Neptune, have $T_{\rm eff} = 59$ K. Jupiter has $T_{\rm eff} = 124$ K, while Earth has $T_{\rm eff} \sim 255$ K. If we could measure $T_{\rm eff}$ for exoplanets, we would expect the hottest hot Jupiters to have $T_{\rm eff} > 2000$ K.

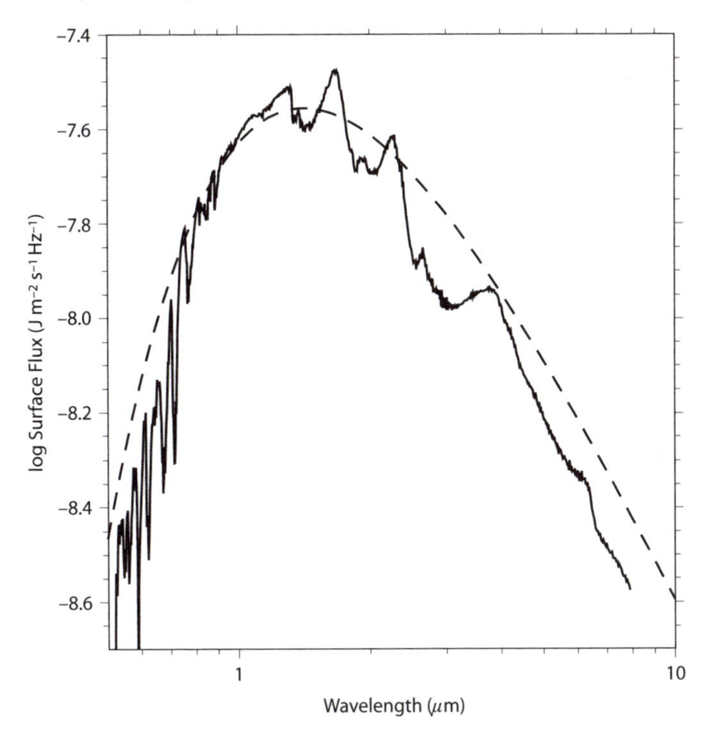

Figure 3.3 Illustration of the effective temperature definition. The theoretical surface flux of a $T_{\rm eff} = 3600$ K M2.5V star is shown [1]. A black body with the same 3600 K effective temperature is shown by the dashed curve. The total black body and stellar flux are equivalent. This star's flux departs from a black body flux; the deep absorption features are due mainly to TiO at visible wavelengths and to H_2O at near-infrared wavelengths.

3.3.2 The Equilibrium Temperature $T_{\rm eq}$

$T_{\rm eq}$ is an estimate of the effective temperature for a planet with no internal luminosity. $T_{\rm eq}$ is a theoretical number and does not refer to any flux measurements of the planet. Physically, $T_{\rm eq}$ is the effective temperature attained by an isothermal planet after it has reached complete equilibrium with the radiation from its parent star. $T_{\rm eq}$ is essential to describe exoplanets, for which it is difficult or impossible to measure the effective temperatures.

$T_{\rm eq}$ can be derived by using the energy balance equation [3.3], that is, by equating the energy emitted by the planet with the energy absorbed by the planet,

$$\frac{4\pi}{f} R_{\rm p}^2 F_{\rm S,p} = (1 - A_{\rm B}) F_{\rm S,*} \left(\frac{R_*}{a} \right)^2 \pi R_{\rm p}^2. \tag{3.7}$$

On the left-hand side we have introduced a parameter f, a correction factor to the term 4π. We use f to enable us to describe the $T_{\rm eq}$ from only one hemisphere of the planet; we shall discuss its meaning and a choice of values later in this subsection.

To derive $T_{\rm eq}$ from energy balance we rewrite equation [3.7], using the Stefan-Boltzmann Law (equation [3.6])

$$F_{\rm S} = \sigma_{\rm R} T_{\rm eff}^4, \tag{3.8}$$

to find

$$\boxed{T_{\rm eq} = T_{\rm eff,*} \left(\frac{R_*}{a}\right)^{1/2} [f'(1 - A_{\rm B})]^{1/4}.} \tag{3.9}$$

Here we have substituted $T_{\rm eq}$ for the planetary $T_{\rm eff}$. We have also absorbed a factor of 1/4 into f' by defining $f' = f/4$.

Remember that for exoplanets we can observe only one hemisphere at a time, and this is what drives our need for the correction factor f. If the absorbed stellar radiation is uniformly redistributed around the planet, that is, into 4π, then $f' = 1/4$. In this case $T_{\rm eq}$ is the same for any hemispheric view of the planet—and indeed the f' correction is not really needed. f' is especially relevant for cases where the absorbed stellar radiation is not redistributed uniformly over the planet. For a slowly rotating or tidally locked planet, the stellar energy is absorbed only by one hemisphere of the planet. If the atmosphere instantaneously reradiates the absorbed radiation (with no advection), $f' = 2/3$. In this case $T_{\rm eq}$ of the day-side hemisphere is much hotter than in the uniform reradiation case.

Beyond the two end cases of uniform redistribution and instantaneous reradiation of absorbed stellar energy, there is no simple or analytical form for f'. If we want to estimate the day-side hemispherical temperature, we must use observations or rely on atmospheric circulation models.

We conclude our discussion of f' with the cautionary remark that f has many different definitions in the literature; we emphasize that our choice of $f' = f/4$ as a correction factor to 4π is deliberate, for symmetry with the albedo derivation for scattered light from the dayside of the planet (Section 3.4.3).

How well does the equilibrium temperature $T_{\rm eq}$ approximate $T_{\rm eff}$ for the solar system planets? Using the planets' measured $A_{\rm B}$s, known semimajor axes, and $f' = 1/4$, we see from Table 3.1 that, with the exception of Uranus, the two temperatures agree to within a few degrees K.

3.3.3 The Brightness Temperature $T_{\rm b}(\nu)$

The flux of exoplanets at different wavelengths or frequencies is the quantity we can measure at Earth. We have called this $\mathcal{F}_\oplus$, where $\mathcal{F}_\oplus = \frac{R_p^2}{D_\oplus^2} F_{\rm S}$ for a very distant planet with surface flux $F_{\rm S}$. The intuitive familiarity we have with temperature makes it easier to use brightness temperature instead of flux to describe exoplanet atmosphere measurements. Moreover, temperatures are more directly relevant than fluxes for the physics and chemistry of exoplanet atmospheres. Fluxes depend on the planet's distance from Earth and so change from one planet to the next, whereas the temperature of a planet atmosphere is distance independent. Measured exoplanet fluxes are therefore often stated as temperatures.

The brightness temperature $T_{\rm b}(\nu)$ is defined as the temperature of a black body of the same shape and at the same distance as the planet and with the same flux as

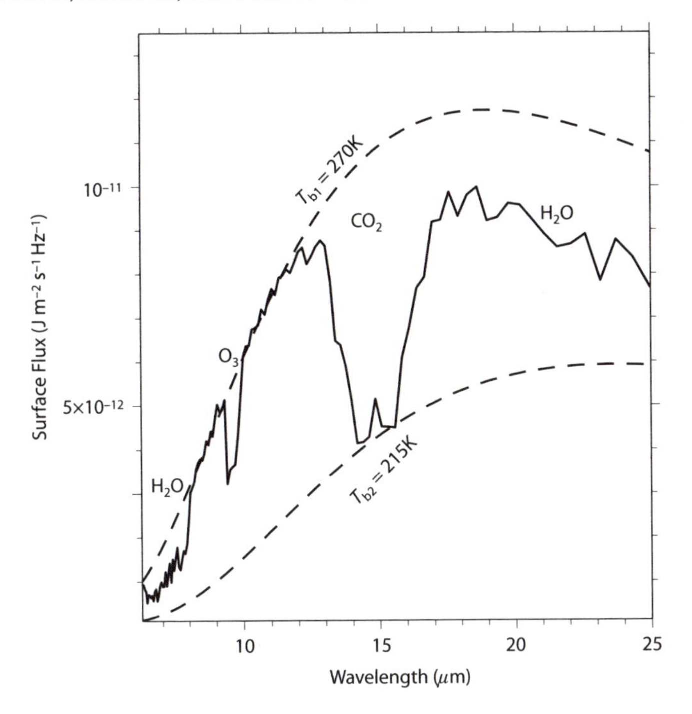

Figure 3.4 Illustration of the brightness temperature definition. The solid curve is Earth's surface flux as measured by the *Mars Global Surveyor* [2]. The dashed curves are black body fluxes with different temperatures, showing that the brightness temperature $T_{\rm b}$ can vary with wavelength or frequency. From Earth's spectrum we see that $T_{\rm b}(12\ \mu{\rm m}) \simeq 270$ K, in contrast to $T_{\rm b}(14.2\ \mu{\rm m}) \simeq 215$ K.

the planet *in a specified frequency range*. $T_{\rm b}(\nu)$ is a frequency-dependent expression of planetary flux,

$$\boxed{\sigma_{\rm R} T_{\rm b}^4(\nu) = F_{\rm S}(\nu) = \mathcal{F}_{\oplus} \left(\frac{D_{\oplus}}{R_p}\right)^2.} \tag{3.10}$$

Here the Stefan-Boltzmann Law (equation [3.6]) is used to convert flux to a temperature. We again note that the brightness temperature assumes the planet behaves as a black body only in the frequency range where the flux is measured. While a black body radiator has a constant $T_{\rm b}(\nu)$, planets with spectra that depart from a black body have $T_{\rm b}(\nu)$ that can vary widely with wavelength or frequency (see Figure 3.4).

We emphasize that $T_{\rm b}(\nu)$ is the only temperature we can currently measure for exoplanets, because for $T_{\rm b}(\nu)$ flux at only one frequency or wavelength (or frequency or wavelength range) need be observed. To get from a measured $T_{\rm b}(\nu)$ to the more global $T_{\rm eff}$, we must resort to a model atmosphere computer code.

Table 3.1 Temperatures and albedos of solar system planets.

Planet	$T_{\rm eff}$	$T_{\rm eq}$	$A_{\rm g}$	$A_{\rm B}$
Mercury			0.106	0.119
Venus	~ 230	230	0.65	0.750
Earth	~ 255	253	0.367	0.306
Mars	~ 212	209	0.150	0.250
Jupiter	124.4 ± 0.3	109	0.52	0.343
Saturn	95.0 ± 0.4	80	0.47	0.342
Uranus	59.1 ± 0.3	58	0.51	0.300
Neptune	59.3 ± 0.8	46	0.42	0.290

$A_{\rm g}(\nu)$ is at ~500 nm. Values from [3].

3.4 PLANETARY ALBEDOS

The planetary albedo is a measure of the reflectivity of the planet's surface and/or atmosphere. More specifically, the albedo is the ratio of the light scattered by a planet to the light received by the planet. Just as for temperature, there is no single albedo to describe the planet atmosphere or surface. Yet the albedo is an important planet parameter since the planet reflectivity is indicative of cloud or surface conditions. Moreover, the albedo controls the planet's energy balance (equation [3.9]) and its effective temperature. Here we describe four commonly used albedos and the relationship among them: the single-scattering albedo, the geometric albedo, the Bond albedo, and the spherical albedo. We also describe the apparent albedo, a very useful albedo quantity for exoplanets. What makes a planet bright or dark? A planet with a high fractional coverage of very reflective clouds is bright, whereas a planet with no clouds or no atmosphere is typically dark. Venus, for example, is the brightest known planet, scattering 0.75 of the incident energy. This is due to its complete (i.e., 100%) coverage of H_2SO_4 clouds. Earth has highly reflective water liquid or ice clouds, but they cover only about 50% of the surface, and as a consequence Earth scatters only 0.3 of the incident energy. Icy bodies may also be bright, with young ice being more reflective than old ice. Notably, Enceladus, a Saturnian satellite, has water geysers which produce fresh ice, making the satellite very bright. Mercury, in contrast, has no atmosphere, clouds, or ice and is dark, scattering only 0.1 of the incident radiation.

3.4.1 Single Scattering Albedo

The single scattering albedo $\tilde{\omega}$ is the fraction of incident light that is scattered by a given particle in the planetary atmosphere. As an example, the single scattering albedo of a water ice crystal in Earth's atmosphere can be as high as 0.8, while that of a plant leaf at visible wavelengths is closer to 0.1. Single scattering albedos are wavelength dependent. The single scattering albedos of Earth's atmosphere and surface vary widely. At the high end, old to fresh snow single scattering albedos

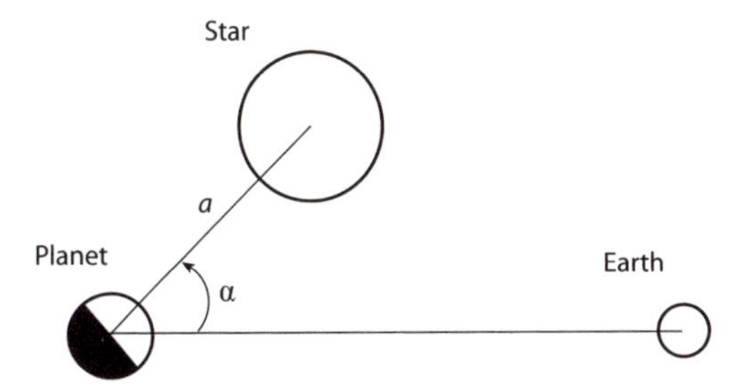

Figure 3.5 Definition of phase angle α. The phase angle is the star-planet-observer angle. Also shown is a, the planet's semimajor axis.

range from ~0.45 to 0.85, and from about 0.35 to 0.8 for clouds. Less reflective, desert sand has a single scattering albedo ranging from about 0.2 to 0.35. At the low reflectivity range is water, whereby oceans have a single scattering albedo less than 0.05.

The single scattering albedo is not a global albedo of the planet atmosphere or surface. Nevertheless, the reflective properties of individual gas, cloud, and surface particles ultimately cause the planet's global albedo through complex multiple scattering of incident radiation.

Important to the overall planet albedo is the directional scattering properties of any individual particle. Few particles in nature, if any, scatter radiation isotropically. Some particles such as ice crystals can have a severe tendency to backscatter incident radiation. This directional scattering property is called the "bidirectional reflection distribution function," often referred to as simply BRDF.

3.4.2 The Phase Angle

Before describing the planetary albedos we first define the planetary phase angle. The phase angle α is the star-planet-observer angle (Figure 3.5). With this definition, $\alpha = 0°$ corresponds to full phase, $\alpha > 170°$ corresponds to a thin crescent phase, and the planet is not at all illuminated at $\alpha = 180°$.

For solar system planets, only Mercury and Venus show all illumination phases as seen from Earth. Outer giant planets show only a few to several degrees in phase angle as seen from Earth. The full phase, or $\alpha = 0$, is a natural configuration for observing solar system planets when the Sun, Earth, and planet are aligned.

For exoplanets, the range of visible planetary phases depends on the orbital inclination of the planet-star system. For example, a planet in a system with orbital inclination of 90° will show all phases as seen from Earth (except phases right near $\alpha = 0$ when the planet is directly behind the star). In contrast, a planet with a "face-on" orbital inclination will have a constant, "half" phase. In contrast to solar system planets, not all exoplanets will be visible at $\alpha \approx 0$; because of the randomness of exoplanet orbital inclinations, $\alpha \approx 0$ may not occur for some planets. In reality, $\alpha \equiv 0$ does not occur for exoplanets because the planet would be directly behind its parent star and not even observable; however, we consider a small α close enough to $\alpha = 0$ for our purposes.

3.4.3 The Geometric Albedo $A_g(\nu)$

3.4.3.1 The Definition of $A_g(\nu)$

The geometric albedo $A_g(\nu)$ is the ratio of a planet's flux at zero phase angle (i.e., full phase) to the flux from a Lambert disk at the same distance and with the same cross-sectional area as the planet (i.e., a surface that subtends the same solid angle). This definition is historical and comes from early observations of solar system planets. To measure the solar system planets' albedo, photographic brightnesses were compared to a photograph of a uniformly illuminated Lambert disk. This relative measurement helped to eliminate instrumental errors. We emphasize that $A_g(\nu)$ is *not* the fraction of incident energy scattered at $\alpha = 0$ (see exercise 3.5).

We derive $A_g(\nu)$ by first considering the numerator in the $A_g(\nu)$ definition: the flux from a planet at zero phase angle as observed at Earth. Choosing a spherical polar coordinate system with latitude θ and longitude ϕ, and considering the flux at Earth $\mathcal{F}_\oplus$ (Chapter 2) we obtain for the flux

$$\mathcal{F}_\oplus(\nu) = \left(\frac{R_p}{D_\oplus}\right)^2 \int_{-\pi/2}^{\pi/2} \int_{-\pi/2}^{\pi/2} I_{S,\mathrm{scat}}(\theta, \phi, \nu) \cos\phi \cos^2\theta d\theta d\phi. \tag{3.11}$$

We have used a subscript "scat" on I to indicate that we are considering only scattered radiation (and not radiation that has been absorbed and thermally reradiated). Recall that the subscript S on I indicates flux at the planet's surface.

Next in the geometric albedo derivation we consider the denominator in the $A_g(\nu)$ definition: the flux from a Lambert disk at the same location and of the same cross-sectional area as the planet. We use the equation for flux observed at Earth ($\mathcal{F}$, equation [3.11]) with the surface element of a disk, $r^2 d\theta$. In writing the flux from a Lambert disk, we also relate the emergent intensity to the incoming intensity by $I_{S,\mathrm{scat}}(\theta, \nu) = I_{\mathrm{inc}}(\theta, \nu)$. This relationship comes from the definition of a Lambert surface, in short that intensity is scattered isotropically (Section 2.9). We further assume that the incident intensity is uniform, that is, $I_{\mathrm{inc}}(\theta, \nu) = I_{\mathrm{inc}}(\nu)$. We then have

$$\mathcal{F}_{\mathrm{L\ disk}}(\nu) = \left(\frac{R_p}{D_\oplus}\right)^2 \int_0^\pi I_{\mathrm{inc}}(\nu) d\theta = \left(\frac{R_p}{D_\oplus}\right)^2 \pi I_{\mathrm{inc}}(\nu). \tag{3.12}$$

Now using the definition of A_g—the ratio of the flux from a planet at phase angle zero to the flux from a Lambert disk, that is, the ratio of equations [3.11] and [3.12]—we find

$$\boxed{A_g(\nu) = \frac{\int_{-\pi/2}^{\pi/2} \int_{-\pi/2}^{\pi/2} I_{S,\mathrm{scat}}(\theta, \phi, \nu) \cos\phi \cos^2\theta d\theta d\phi}{\pi I_{\mathrm{inc}}(\nu)}.} \tag{3.13}$$

We notice something interesting about this expression for $A_g(\nu)$: the denominator can be written as $\pi I_{\mathrm{inc}}(\nu) = \mathcal{F}_{\mathrm{inc}}(\nu)$, the incident flux from the star at the planet's substellar point. So we may also describe $A_g(\nu)$ by the ratio of the planet's scattered flux to the ratio of the incident stellar flux at the substellar point.

We pause here to further point out that equation [3.13] is the form to use to compute $A_g(\nu)$ from an exoplanet model atmosphere code output. The model atmosphere code will generate the scattered intensity at the planetary surface as a function of location on the planet surface (here denoted by θ and ϕ).

We now continue to simplify equation [3.13] for a more conceptual understanding. We can rewrite the geometric albedo considering the relationship between the incoming intensity ($I_{\rm inc}$) and the intensity scattered out by the planet ($I_{\rm S,scat}$). We do this by introducing the dimensionless term $p(\Theta)$: the fraction of incident intensity scattered out of the planet into angle Θ. Here $\Theta = 0$ is defined in the direction of the incident intensity (see Figure 5.1). In this description, all of the physics of directional scattering and multiple scattering inside the planet atmosphere is buried in the final form of $p(\Theta)$,

$$I_{\rm S,scat}(\theta, \phi, \nu) = I_{\rm inc}(\theta', \phi', \nu) p(\Theta). \tag{3.14}$$

The terms with primes refer to the direction of incidence and the terms without primes refer to the direction after scattering. We digress to mention that $p(\Theta)$ here is not the same as the dimensionless single scattering phase function $P(\Theta)$ used in later chapters.

We can further simplify the above intensity relationship by assuming that the incident stellar intensity is uniform and noting that the incident stellar intensity is diluted away from the substellar point as

$$I_{\rm inc}(\theta', \phi', \nu) = \hat{\mathbf{n}} \cdot \hat{\mathbf{k}} I_{\rm inc}(\theta'_0, \phi'_0, \nu) = \cos\Theta I_{\rm inc}(\nu). \tag{3.15}$$

See Figure 2.3 to see that $\hat{\mathbf{n}} \cdot \hat{\mathbf{k}} = \cos\Theta$. From Figure 2.3 and exercise 2.4, we also have

$$\cos\Theta = \cos\theta \cos\phi, \tag{3.16}$$

and therefore

$$I_{\rm S,scat}(\theta, \phi, \nu) = \cos\theta' \cos\phi' I_{\rm inc}(\nu) p(\theta, \phi, \theta', \phi'). \tag{3.17}$$

The A_g from equation [3.13] then becomes

$$A_g(\nu) = \frac{\int_{-\pi/2}^{\pi/2} \int_{-\pi/2}^{\pi/2} I_{\rm inc}(\nu) p(\theta, \phi, \theta', \phi') \cos^2\phi' \cos^3\theta' d\theta' d\phi'}{\pi I_{\rm inc}(\nu)}. \tag{3.18}$$

We may cancel out the $I_{\rm inc}(\nu)$ in the numerator and denominator to find

$$\boxed{A_g(\nu) = \frac{1}{\pi} \int_{-\pi/2}^{\pi/2} \int_{-\pi/2}^{\pi/2} p(\theta, \phi, \theta', \phi') \cos^2\phi' \cos^3\theta' d\theta' d\phi'.} \tag{3.19}$$

3.4.3.2 *The A_g for a Lambert Sphere*

As an example of geometric albedo we will calculate the $A_g(\nu)$ for a Lambert sphere. Recall that a Lambert surface scatters intensity equally in all directions. A Lambert sphere therefore has

$$p(\Theta) = 1. \tag{3.20}$$

Using the A_g definition in equation [3.19],

$$A_{g\text{Lambert}}(\nu) = \frac{1}{\pi}\int_{-\pi/2}^{\pi/2}\int_{-\pi/2}^{\pi/2}\cos^2\phi\cos^3\theta d\theta d\phi = \frac{2}{3}. \tag{3.21}$$

The geometric albedo for a Lambert sphere is $A_g(\nu) = 2/3 < 1$, even though the definition of a Lambert sphere is that all incident radiation is scattered and none is absorbed. This means that for a Lambert sphere one-third of the incident radiation is scattered out of the line of sight.

By way of the $A_g(\nu)$ definition, the geometric albedo may be greater than 1. This is because the geometric albedo is not the fraction of incident radiation scattered at $\alpha = 0$, but the scattered incident radiation compared to the scattered incident radiation from a perfectly diffusing disk at the same location and of the same size. Saturn's icy moon Enceldaus has $A_g = 1.38$ at visible wavelengths [4]. Ice can have a strong backscattering effect, making the scattered radiation greater than the scattered radiation from a diffusing disk.

Measurements of A_g have been elusive for exoplanets. Nevertheless, A_g is still useful for classifying theoretical models.

3.4.4 The Bond Albedo A_B and the Spherical Albedo A_S

A_B is the fraction of incident stellar energy scattered back into space by the planet. A_B includes the radiation scattered into all directions and radiation at all frequencies. By its definition, $A_B \leq 1$.

The spherical albedo $A_S(\nu)$ has the same definition as the Bond albedo but at a specific frequency,

$$A_B = \int_0^\infty A_S(\nu)d\nu. \tag{3.22}$$

We emphasize up front that the Bond albedo, through the definition of the spherical albedo below, is actually weighted by the incident radiation. Both the Bond and spherical albedos therefore depend on the spectrum of the planet's host star.

To derive $A_s(\nu)$ we begin by revisiting the numerator in $A_g(\nu)$ (equation [3.13]), the scattered radiation from the planet at $\alpha = 0$. Recall that the $A_g(\nu)$ numerator is an expression for flux (equation [3.11]). The important point to keep in mind is that at each phase angle α only a portion of the hemisphere facing Earth is illuminated. This is illustrated in Figure 3.6. Using Figure 3.6 we construct a more general form of the light scattered from the planet surface,

$$F_{S,\alpha}(\nu) = \int_{\alpha-\pi/2}^{\pi/2}\int_{-\pi/2}^{\pi/2} I_{\text{inc}}(\theta',\phi',\nu)p(\theta,\phi,\theta',\phi')\cos(\alpha-\phi')\cos\phi'\cos^3\theta' d\theta' d\phi' \tag{3.23}$$

and we make the assumption that the incident intensity is uniform to write

$$F_{S,\alpha}(\nu) = I_{\text{inc}}(\nu)\Psi_\alpha(\nu). \tag{3.24}$$

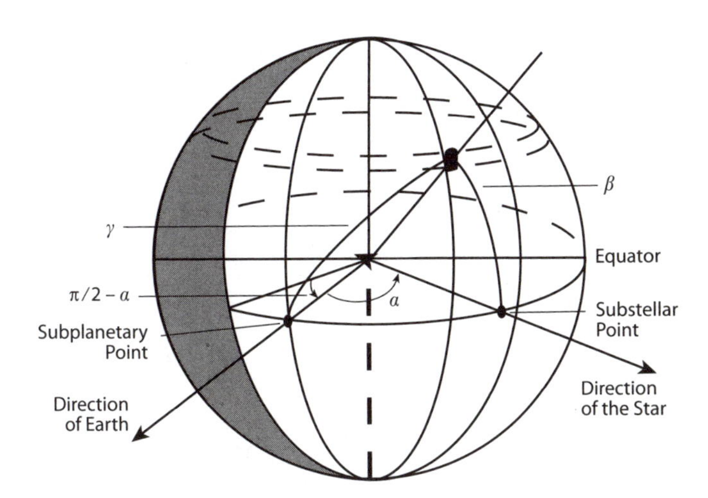

Figure 3.6 Spherical planetary coordinates suitable for deriving the Lambert sphere phase curve. θ is the latitude measured from the intensity equator. ϕ is the longitude measured from the sub-Earth point. α is the phase angle. After [5].

Here Ψ_α is the fraction of scattered radiation in the *direction of the observer* as a function of α.

To account for the total scattered energy in *all directions* we must integrate $F_{S,\alpha}(\nu)$ over all directions. We want to know the total energy per unit time per frequency scattered by the planet compared to the total energy per unit time per frequency incident on the planet. We consider a sphere around the planet with a spherical polar coordinate system. We adopt the angles α as our altitude and β as our azithmuthal angle. We further assume azimuthal symmetry. To find the total scattered energy in all directions, we integrate $F_{S,\alpha}(\nu)$ (defined in equation [3.24]) over $(0 \leq \alpha < \pi)$ and $(0 \leq \beta < 2\pi)$,

$$E_{\rm scat}(\nu) = I_{\rm inc}(\nu) \int_0^{2\pi} \int_0^{\pi} \Psi_\alpha(\nu) R_{\rm p}^2 \sin\alpha d\alpha d\beta. \tag{3.25}$$

and because of the azithmuthal symmetry,

$$E_{\rm scat}(\nu) = 2\pi R_{\rm p}^2 I_{\rm inc}(\nu) \int_0^{\pi} \Psi_\alpha(\nu) \sin\alpha d\alpha. \tag{3.26}$$

We now consider the total incident energy on the planet as related to equation [2.31],

$$E_{\rm inc}(\nu) = \pi R_{\rm p}^2 \pi I_{\rm inc}(\nu). \tag{3.27}$$

The spherical albedo is the ratio of scattered energy to incident energy

$$A_{\rm S}(\nu) = \frac{E_{\rm scat}(\nu)}{E_{\rm inc}(\nu)} = \frac{2 I_{\rm inc}(\nu) \int_0^{\pi} \Psi_\alpha(\nu) \sin\alpha d\alpha}{\pi I_{\rm inc}(\nu)}, \tag{3.28}$$

$$\boxed{A_\mathrm{S}(\nu) = \frac{2}{\pi}\int_0^\pi \Psi_\alpha(\nu)\sin\alpha d\alpha.} \tag{3.29}$$

3.4.5 The Relationship between the Bond and Geometric Albedos

In order to measure the Bond or spherical albedo we would need to observe the planet at all phase angles. All phase angles are not usually visible from Earth. In order to estimate the spherical albedo, it is useful to relate the spherical albedo to the geometric albedo. We have from the above equations [3.19] and [3.23]

$$\Psi_{\alpha=0}(\nu) \equiv \frac{A_\mathrm{g}(\nu)}{\pi}. \tag{3.30}$$

We may therefore redefine the spherical albedo $A_\mathrm{S}(\nu)$ in equation [3.29] as

$$A_\mathrm{S}(\nu) = \frac{\pi A_\mathrm{g}(\nu)}{\Psi_{\alpha=0}(\nu)}\frac{2}{\pi}\int_0^\pi \Psi_\alpha(\nu)\sin\alpha d\alpha, \tag{3.31}$$

which we can also summarize as

$$\boxed{A_\mathrm{s}(\nu) = A_\mathrm{g}(\nu)q(\nu),} \tag{3.32}$$

where $q(\nu)$ is the phase integral. The phase integral is formally defined as

$$q(\nu) \equiv \frac{A_\mathrm{s}(\nu)}{A_\mathrm{g}(\nu)} = 2\int_0^\pi \frac{\Psi_\alpha(\nu)}{\Psi_{\alpha=0}(\nu)}\sin\alpha d\alpha. \tag{3.33}$$

We can more conveniently write

$$\boxed{q(\nu) = 2\int_0^\pi \Phi_\alpha(\nu)\sin\alpha d\alpha.} \tag{3.34}$$

$\Phi_\alpha(\nu)$ is the phase function: the fractional scattered flux variation with phase angle normalized to the flux at $\alpha = 0$,

$$\boxed{\Phi_\alpha(\nu) = \frac{\Psi_\alpha(\nu)}{\Psi_{\alpha=0}(\nu)}.} \tag{3.35}$$

The spherical albedo can be estimated from equation [3.33] with both a measured $A_\mathrm{g}(\nu)$ and a theoretical calculation for $q(\nu)$.

What is the relationship between the geometric and spherical albedos for a Lambert sphere? For a Lambert surface, $\Phi_\alpha(\nu) = \cos\alpha$ and using equations [3.32] and [3.33] we find

$$A_\mathrm{g}(\nu) = \frac{2}{3}A_\mathrm{s}(\nu). \tag{3.36}$$

For a Lambert sphere the geometric albedo is lower than the spherical albedo by a factor of 2/3. We can understand this conceptually as follows. For a Lambert sphere all radiation is scattered back to space and, since the spherical albedo includes radiation into all angles $A_s(\nu) \equiv A_\mathrm{B} = 1$. The geometric albedo, in contrast is the albedo only at phase angle 0 and includes only the radiation backscattered to the observer, so it is less than 1.

A comparison of Bond and total geometric albedos of Jupiter, Saturn, Uranus, and Neptune is shown in Figure 3.7. Here we call "total geometric albedo" the geometric albedo integrated over all wavelengths. We can see that these bodies with deep atmospheres all lie above the line for isotropic scattering (i.e., a Lambert sphere).

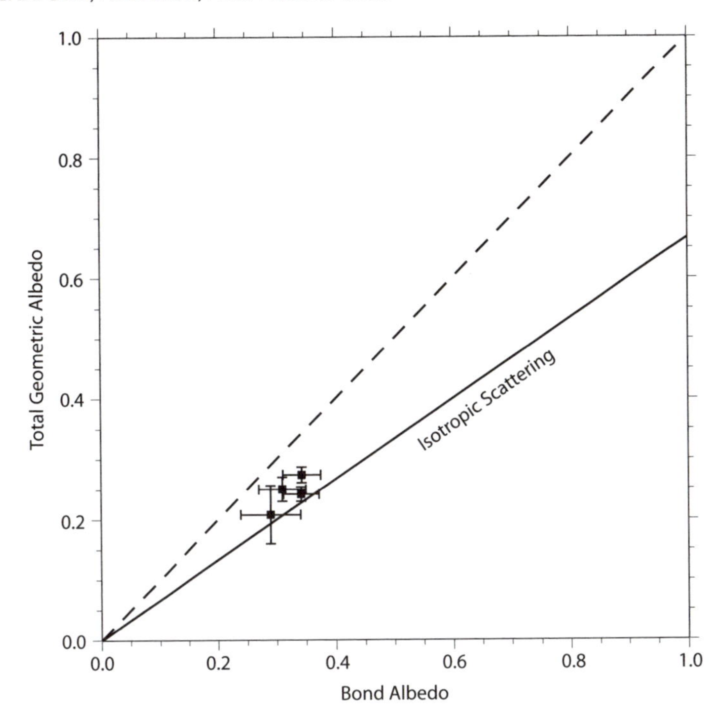

Figure 3.7 Relationship between the Bond albedo (A_B) and geometric albedo (A_g) for a Lambert sphere and solar system giant planets. Here we use the "total" A_{gTot} to refer to A_g over all visible wavelengths. The points are for Uranus, Neptune, Saturn, and Jupiter (in order of increasing A_B). The solid line ($A_{gTot}/A_B = 0.67$) is for Lambertian isotropic reflectance (i.e., constant for all angles of incidence). The dashed line is the line of equivalence where $A_{gTot} = A_B$ (all gas giant planets with deep atmospheres must lie to its right). The wedge between the solid and dashed lines defines a useful limiting region; it bounds the photometric properties of most spherical bodies with deep atmospheres (with, e.g., Rayleigh scattering, clouds, dust, etc.). Figure adapted from [6]. Solar system data points are referenced in [7].

3.4.6 The Apparent Albedo

The geometric, spherical, and Bond albedos are not very satisfactory for exoplanets because none can be measured. The geometric albedo is defined at full phase ($\alpha = 0$), a configuration where the planet is behind the star and not observable. The spherical and Bond albedos are also not measurable because all phase angles are not accessible for exoplanets. With the direct imaging technique, many phase angles where the planet is projected too close to the star will not be observeable. Moreover, unless the planet orbit is edge-on to our line of sight the planet will not go through all phases as seen from Earth. We therefore turn to an albedo definition that is more appropriate for exoplanets.

The apparent albedo can be defined as the ratio between the scattered flux emerging from the planet in the direction of the observer and the scattered flux from a perfectly reflecting Lambert sphere with the same size as the planet, at the same phase angle, and at the same distance from the observer. The apparent albedo can be understood as a planet's reflectance in the direction of the observer at a given time. The apparent albedo is the only albedo quantity that can potentially be derived from a single observation, for a planet of known size. The apparent albedo is a function of the phase angle and frequency. We leave it as an exercise to formulate an equation for the apparent albedo.

3.5 PLANET-STAR FLUX RATIOS

An estimate of the flux ratio of the planet and star is useful for exploring whether a particular planet and star combination can be detected with a given telescope. For the planet-star flux ratio estimate, we assume that the stars and planets radiate as black bodies. Figure 3.8 shows planet-star flux ratios for the Sun and solar system planets, as well as for a hypothetical hot Jupiter. The planets have two peaks in their spectra. The short-wavelength peak is due to starlight scattered from the planet atmosphere, and therefore has a flux peak at the same wavelength as the star. For each planet, the long-wavelength flux peak is from the planet's thermal emission and peaks at the planet's characteristic temperature. The high contrast at all wavelengths of the planet-to-star flux ratio is the primary challenge for detecting radiation from exoplanets.

We recall that the planet flux at Earth is (equation [2.19])

$$\mathcal{F}_{\oplus,p} = F_{\mathrm{S,p}} \left(\frac{R_{\mathrm{p}}}{D_{\oplus}} \right)^2 , \tag{3.37}$$

where the subscript $\oplus$ refers to a flux measurement at Earth and the subscript S refers to flux at the planet's surface. Similarly, the star's flux at Earth is

$$\mathcal{F}_{\oplus,*} = F_{\mathrm{S},*} \left(\frac{R_*}{D_{\oplus}} \right)^2 , \tag{3.38}$$

To find the planet-star flux ratio, we divide the planet flux at Earth by the star flux at Earth,

$$\boxed{\frac{\mathcal{F}_{\oplus,\mathrm{p}}}{\mathcal{F}_{\oplus,*}} = \frac{F_{\mathrm{S,p}} R_{\mathrm{p}}^2}{F_{\mathrm{S},*} R_*^2}.} \tag{3.39}$$

Note that we can use a similar equation for frequency-dependent flux.

3.5.1 Thermal Emission Flux Ratio

We can estimate the thermal emission planet-star flux ratio on the illuminated side of the planet by equating the planet's thermal emission with the absorbed stellar radiation, in a rearranged form of energy balance described in equation [3.7],

$$\boxed{\frac{\mathcal{F}_{\oplus,\mathrm{p}}}{\mathcal{F}_{\oplus,*}} = \frac{F_{\mathrm{S,p}} R_{\mathrm{p}}^2}{F_{\mathrm{S},*} R_*^2} = \left(\frac{R_{\mathrm{p}}}{a} \right)^2 \frac{f}{4} (1 - A_{\mathrm{B}}) .} \tag{3.40}$$

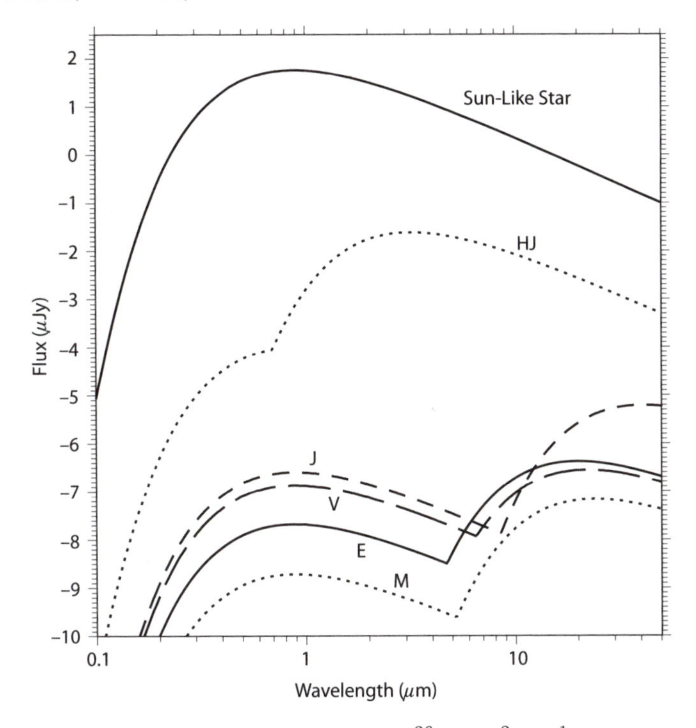

Figure 3.8 The approximate spectra (in units of 10^{-26} W m^{-2} Hz^{-1}) of some solar system bodies as seen from a distance of 33 light years. The Sun is represented by a black body of 5750 K. The planets Jupiter, Venus, Earth, and Mars are shown and are labeled with their first initial. A representative hot Jupiter exoplanet is also shown.

Here we recall that f is a correction factor to 4π to describe the unknown redistribution of absorbed stellar radiation. We further note that the fluxes in the above equation are total fluxes, integrated over all frequencies.

A more useful planet-star flux ratio estimate is one that is frequency dependent, because telescope observations of exoplanets are limited to a fixed frequency or wavelength range. We may approximate the star and planet surface fluxes with black body fluxes $\pi B(T, \nu)$, where T is the effective temperature of the star or the equilibrium effective temperature of the planet,

$$\frac{\mathcal{F}_{\oplus,\mathrm{p}}(\nu)}{\mathcal{F}_{\oplus,*}(\nu)} = \frac{F_{\mathrm{S,p}}(\nu) R_{\mathrm{p}}^2}{F_{\mathrm{S},*}(\nu) R_*^2} = \left[\frac{\mathrm{e}^{h\nu/kT_{T_{\mathrm{eff},*}}} - 1}{\mathrm{e}^{h\nu/kT_{\mathrm{eq,p}}} - 1}\right] \left(\frac{R_{\mathrm{p}}^2}{R_*^2}\right). \tag{3.41}$$

At mid-infrared wavelengths we may make a further simplification for the planet-star flux ratio. This simplification is based on the point that the black body fluxes for reasonable planet and star temperatures are both in the Rayleigh-Jeans region of the black body spectrum, $h\nu/kT \ll 1$. Under this approximation, $\mathrm{e}^{-x} \approx 1 - x$, and the black body flux (using equation [2.34]) becomes

$$\pi B(T, \nu) \approx \pi \frac{2\nu^2}{c^2} kT, \tag{3.42}$$

and the flux ratio becomes

$$\boxed{\frac{\mathcal{F}_{\oplus,\mathrm{p}}(\nu)}{\mathcal{F}_{\oplus,*}(\nu)} = \frac{F_{\mathrm{S,p}}(\nu)R_{\mathrm{p}}^2}{F_{\mathrm{S},*}(\nu)R_*^2} = \left[\frac{T_{\mathrm{eq,p}}}{T_{\mathrm{eff},*}}\right]\left(\frac{R_{\mathrm{p}}}{R_*}\right)^2} \tag{3.43}$$

in the frequency range where the Rayleigh-Jeans approximation is valid.

The planet-star flux ratio in equation [3.43] is for planet and star atmospheres represented by black body fluxes. In acknowledgement that planet and star fluxes can deviate significantly from those of black bodies, we should use the brightness temperature $T_{\mathrm{b}}(\nu)$ at individual frequencies. $T_{\mathrm{b}}(\nu)$ can be estimated from model atmosphere calculations. Using an earlier version of the planet-star flux ratio (from the left-hand side of equation [3.40]), we can rewrite equation [3.43] at individual frequencies, using the brightness temperature,

$$\frac{\mathcal{F}_{\oplus,\mathrm{p}}(\nu)}{\mathcal{F}_{\oplus,*}(\nu)} = \frac{F_{\mathrm{S,p}}(\nu)R_{\mathrm{p}}^2}{F_{\mathrm{S},*}(\nu)R_*^2} = \left[\frac{T_{\mathrm{b,p}}(\nu)}{T_{\mathrm{b},*}(\nu)}\right]\left(\frac{R_{\mathrm{p}}}{R_*}\right)^2. \tag{3.44}$$

3.5.2 Scattered Light Flux Ratio

The planet-star flux ratio in terms of the scattered radiation by the planet is different from the planet-star flux ratio for thermal radiation emitted by the planet. The planet's scattered radiation is typically at visible wavelengths but may extend to much longer wavelengths for colder planets (see Figure 3.8). In this subsection we will concern ourselves with the planet-star flux ratio at $\alpha = 0$.

We again start with the planet-star flux ratio equation [3.40]. Because we are now interested in the scattered stellar flux instead of the absorbed stellar flux, we replace the term that describes the absorbed and reradiated stellar radiation on the illuminated hemisphere,

$$\frac{f}{4}\left(1 - A_{\mathrm{B}}\right), \tag{3.45}$$

with a term to describe the scattered radiation on the illuminated hemisphere,

$$\frac{g}{4}A_s(\nu). \tag{3.46}$$

We have used the variable g to replace f. Like f, g is a correction factor to 4π—specifically, g describes the solid angle into which the incident stellar radiation is reradiated, compared to 4π. To explain the need for g in more detail, consider that one hemisphere of the planet is illuminated by the star. Is the incident radiation scattered equally into $4\pi R_{\mathrm{p}}^2$? No, because if the reradiation is scattered back toward the observer, it cannot be advected around the planet.

In equation [3.46] we use the spherical albedo $A_{\mathrm{s}}(\nu)$ instead of the Bond albedo A_{B} because we are interested in a wavelength-dependent flux ratio. Finally, for simplicity we assume monochromatic scattering. The planet-star flux ratio at visible wavelengths, including all of the above, is

$$\boxed{\frac{\mathcal{F}_{\mathrm{p},\oplus}(\nu)}{\mathcal{F}_{*,\oplus}(\nu)} = \frac{F_{\mathrm{p,S}}(\nu)R_{\mathrm{p}}^2}{F_{*,\mathrm{S}}(\nu)R_*^2} = \frac{R_{\mathrm{p}}^2}{a^2}\left[\frac{g}{4}\right]A_{\mathrm{s}}(\nu).} \tag{3.47}$$

Many readers may be satisfied to stop here with the definition of the scattered light correction factor (derived below)

$$g = \frac{4A_g(\nu)}{A_s(\nu)}. \tag{3.48}$$

It is a *very common misunderstanding* in this derivation of visible-wavelength flux ratio to not recognize that this flux ratio involves the spherical albedo $A_s(\nu)$—a quantity that describes the fraction of incident energy scattered in all directions. For the planet-star flux ratio at planet full phase ($\alpha = 0$) we need to replace $A_s(\nu)$ with a quantity that describes the scattered energy at full phase, namely, related to $A_g(\nu)$.

With equation [3.48] we may write the visible wavelength planet-star flux ratio as

$$\boxed{\frac{\mathcal{F}_{p,\oplus}(\nu)}{\mathcal{F}_{*,\oplus}(\nu)} = \frac{F_{S,p}(\nu) R_p^2}{F_{S,*}(\nu) R_*^2} = A_g(\nu) \frac{R_p^2}{a^2}.} \tag{3.49}$$

The flux ratio depends inversely on the planet's semimajor axis. Let us consider the visible-wavelength flux ratio of Jupiter (Figure 3.8), considering $A_g(500 \text{ nm}) = 0.5$, $a = 5.2$ AU, and $R_p = R_J$. Using equation [3.49] we find $\mathcal{F}_{p,\oplus}/\mathcal{F}_{*,\oplus} = 4.4 \times 10^{-09}$. Scaling this to a planet at 0.05 AU, the flux ratio is 10^4 times higher, or 4.4×10^{-4}. The actual expected flux ratio of a hot Jupiter orbiting its star at 0.05 AU (or a corresponding few-day period) is not so high compared to Jupiter because Jupiter is bright from its icy clouds of ammonia and water, whereas the hot Jupiters are dark due to lack of bright clouds.

We now return to the correction factor g. This correction factor is the most critical step in the derivation of the scattered light planet-star flux ratio at phase angle zero. g describes the ratio of energy emitted at zero phase angle to energy emitted at all phase angles. From equation [3.29] for scattered energy in all directions and using a form of equation [3.24] for scattered energy in the direction of the observer, we therefore have

$$\frac{g}{4\pi} = \frac{E_{\text{scat},\alpha=0}}{E_{\text{scat}}} = \frac{\Psi_{\alpha=0}(\nu)}{2\pi \int \Psi_\alpha(\nu) \sin \alpha d\alpha}. \tag{3.50}$$

We can use our previous definition of Φ and $q(\nu)$ to find

$$\frac{g}{4\pi} = \frac{1}{\pi q(\nu)} = \frac{A_g(\nu)}{\pi A_s(\nu)} \tag{3.51}$$

and simplify to

$$g = \frac{4A_g(\nu)}{A_s(\nu)}. \tag{3.52}$$

This derivation of g is the final step in finding the scattered light planet-star flux ratio given in equation [3.49].

3.5.3 Transmission Flux Ratio

During an exoplanet transit, stellar radiation is *transmitted* through the planet atmosphere, picking up some spectral features from the planet atmosphere. We call this the transmit transmission spectrum. In practice the planet transit transmission spectrum is not measured alone. Astronomers measure the total flux from the planet and star during transit (the "in-transit flux") and the flux outside of transit (the stellar flux or the "out-of-transit" flux). Dividing these two quantities gives us the in-transit to out-of-transit flux ratio, what we simply call the transmission flux ratio. Here we will estimate the magntiude of the transmission flux ratio.

We begin with the in-transit flux by using the derivation of the planet flux at Earth in Section 2.5 (and equations [2.16] and [2.19]). For the planet, we use $R_{\rm p}$ to mean the radius where the planet is opaque at all wavelengths of interest and A_H to mean the additional radial extent of the planetary atmosphere (in the same units as $R_{\rm p}$).

We consider the angles subtending the star, planet, and planet atmosphere:

$$\begin{aligned}\mathcal{F}_{\text{in trans},\oplus}(\nu) = & \int_0^{2\pi}\int_{(R_{\rm p}+A_H)/D_\oplus}^{R_*/D_\oplus} I_*(\vartheta,\phi,\nu)\cos\omega\sin\omega d\omega d\phi \\ & + \int_0^{2\pi}\int_0^{R_{\rm p}/D_\oplus} I_{\rm p}(\vartheta,\phi,\nu)\cos\omega\sin\omega d\omega d\phi \\ & + \int_0^{2\pi}\int_{R_{\rm p}/D_\oplus}^{(R_{\rm p}+A_H)/D_\oplus} I_{\rm atm}(\vartheta,\phi,\nu)\cos\omega\sin\omega d\omega d\phi. \end{aligned} \quad (3.53)$$

The intensity passing through the opaque part of the planet, $I_{\rm p}$, is zero. We follow Section 2.5 and divide by the observed stellar flux $\mathcal{F}_{*,\oplus}(\nu) = F_{{\rm S},*}\left(R_*/D_\oplus\right)^2$ (from equation [2.19]) to find

$$\boxed{\frac{\mathcal{F}_{\text{in trans},\oplus}(\nu)}{\mathcal{F}_{*,\oplus}(\nu)} = 1 - \left(\frac{R_{\rm p}}{R_*}\right)^2 + \frac{2R_{\rm p}A_H}{R_*^2}\left(1 - \frac{F_{\rm S,p,trans}(\nu)}{F_{{\rm S},*,\rm trans}(\nu)}\right).} \quad (3.54)$$

Here we have used the subscript "in trans, $\oplus$" to refer to the overall flux measured at Earth during transit, the subscript "S, p, trans" to refer to just the flux from the planet atmosphere, and "$S_{*,\rm trans}$" to refer to radiation from the star incident on the planet atmosphere. We have assumed that $A_H \ll R_{\rm p}$, and $F_{\rm S,p,trans}$ is further described in Section 6.4.1. We emphasize that the choice of A_H is not important as long as it is large enough. For example, beyond some atmospheric height A_H, there is no atmosphere left to attenuate the stellar radiation, $F_{\rm S,p,trans}/F_{{\rm S},*,\rm trans} = 1$, and there is no contribution to the in-transit flux.

How can we estimate the magnitude of the transmission flux signal? We will consider the last term on the right-hand side of the above equation—this is the planet-to-star transmission flux ratio. We will estimate the maximum $\mathcal{F}_{\rm S,p,trans}$ as if the atmosphere *annulus* were completely opaque to the stellar radiation, that is, at some wavelengths no radiation is transmitted; all radiation is blocked by the atmosphere, $\mathcal{F}_{\rm S,p,trans} = 0$. We have for the planet-to-star transmission flux ratio

$$\frac{\mathcal{F}_{\rm p,\oplus,trans}(\nu)}{\mathcal{F}_{*,\oplus}(\nu)} = \frac{2R_{\rm p}A_H}{R_*^2}\left(1 - \frac{F_{\rm S,p,trans}(\nu)}{F_{{\rm S},*,\rm trans}(\nu)}\right) \approx \frac{2R_{\rm p}A_H}{R_*^2}. \quad (3.55)$$

As an example, let us consider a short-period Jupiter transiting a sun-sized star. The hot Jupiter is composed primarily of H and He, and with $T_{\rm eq} \simeq 1000$ K the scale height (equation [9.11]) is $H \simeq 500$ km. Typically, we may take the planet atmosphere to be $5 \times H$, that is, $A_H = 5H$. The resulting planet annulus compared to star area is 1.5×10^{-3} (i.e., the last term in the above equation [3.54]). This number may seem very small. In fact, one part per thousand is a relatively large number in the field of exoplanet atmosphere detection. We can see this by comparing to the planet-to-star flux ratios earlier in this section. In contrast to the short-period hot Jupiters, for a colder planet such as a Jupiter analog (a Jupiter-sized planet in a Jupiter-like orbit with $T_{\rm eq} \sim 120$ K) with $H = 24$ km, the estimated transmission spectrum strength is only 7×10^{-5}.

The measurement of an exoplanet spectrum for hot Jupiters has been a successful technique for bright star targets, with sodium, water vapor, methane, and other gases identified in this way.

3.6 PLANETARY PHASE CURVES

An exoplanet goes thorough illumination phases as seen from Earth. These phases are akin to the phases the moon goes through. We are interested not only in the flux ratio at $\alpha = 0$ (i.e., full phase) but also the flux ratio as a function of α. In this section we discuss the planet phase curves, building on previous equations in this chapter.

3.6.1 Visible-Wavelength Phase Curves

We first recall the normalized phase function equation [3.35]:

$$\Phi_\alpha(\nu) = \frac{\Psi_\alpha(\nu)}{\Psi_{\alpha=0}(\nu)}, \tag{3.56}$$

where $\Psi_\alpha(\nu)$ describes the fraction of scattered radiation at any phase angle and is defined in equation [3.23],

$$\Psi(\alpha, \nu) = \int_{\alpha-\pi/2}^{\pi/2} \int_{-\pi/2}^{\pi/2} I_{\rm inc}(\theta', \phi', \nu) p(\theta, \phi, \theta', \phi') \cos(\alpha - \phi') \cos\phi' \cos^3\theta' d\theta' d\phi'. \tag{3.57}$$

There is one case where the phase curve can be analytically derived, that of a Lambert sphere. We use the definition of a Lambert sphere $p(\theta, \phi, \theta', \phi') = 1$ (introduced in section 3.4.3) and assume that the incident radiation is uniform, thus taking $I_{\rm inc}$ out of the integrand in equation [3.57]. We then compute the integral in equation [3.57] normalized as shown in the phase curve definition in equation [3.56] to derive the frequency-independent phase curve of a Lambert sphere

$$\boxed{\Phi_\alpha = \frac{1}{\pi}\left[\sin\alpha + (\pi - \alpha)\cos\alpha\right].} \tag{3.58}$$

The Lambert sphere phase curve is shown in Figure 3.9.

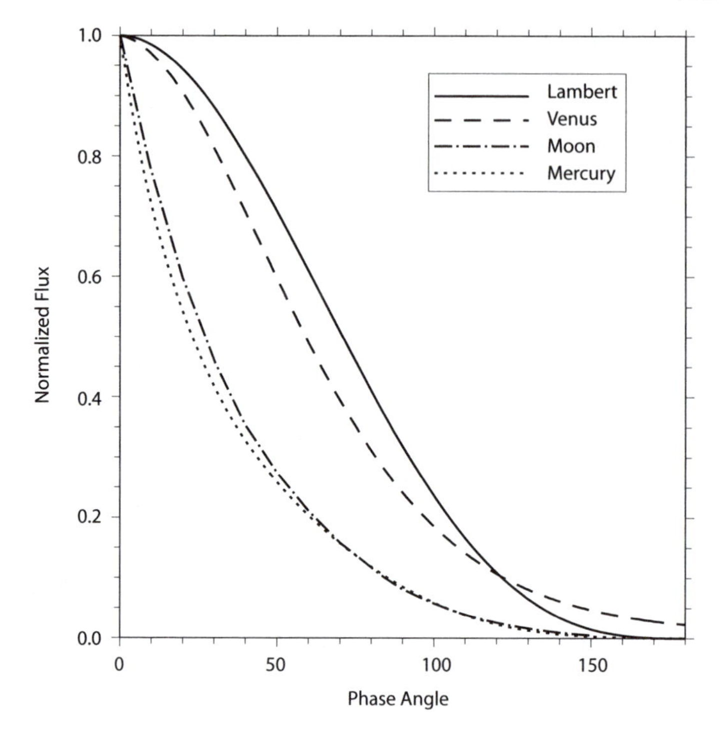

Figure 3.9 Visible-wavelength illumination phase curves for Venus, Mercury, and the Moon. A Lambert sphere phase curve is shown as the rightmost solid black line. From formulae in [3].

Visible-wavelength phase curves of the Moon, Mercury, and Venus are also shown in Figure 3.9. We can see that the atmosphereless bodies Mercury and the Moon have phase curves that are very different from a Lambert sphere.

The phase curve in equation [3.58] is normalized to 1. If we want to compute the planet-star flux ratio at all phases we may use this equation in conjunction with the planet-star flux ratio at $\alpha = 0$ (equation [3.49]):

$$\frac{\mathcal{F}_{\mathrm{p},\oplus,\alpha}(\nu)}{\mathcal{F}_{*,\oplus}(\nu)} = A_{\mathrm{g}}(\nu)\frac{R_{\mathrm{p}}^2}{a^2}\Phi_\alpha(\nu). \tag{3.59}$$

Let us consider the exoplanet example of Earth orbiting the Sun. At $\alpha = 0$, with $A_{\mathrm{g}}(500\text{–}800\ \mathrm{nm}) = 0.2$, $a = 1.0$ AU, and $R_{\mathrm{p}} = R_\oplus$, $\mathcal{F}_{\mathrm{p},\oplus,\alpha}(\nu)/\mathcal{F}_{*,\oplus}(\nu) = 2.1 \times 10^{-10}$. Earth is almost ten billion times fainter than the Sun. Recall that $\alpha = 0$ corresponds to an exoplanet's position directly behind its star as seen from Earth. If we want to know the planet-star flux ratio at the planet-star maximum separation, we must take $\alpha = 90^\circ$. Approximating Earth as a Lambert sphere (equation [3.58]), the planet-star flux ratio is 1.2×10^{-10}, a factor of π lower than the zero-phase-angle value.

3.6.2 IR-Wavelength Phase Curves

An exoplanet may have phase variation at infrared wavelengths from thermal emission, and not just at visible wavelengths due to scattered stellar radiation. Thermal phase variation may arise for an exoplanet whose thermal emission is dominated by reemitted absorbed stellar radiation rather than any internal energy. One class of planets that fits this condition is the tidally locked hot exoplanets. These hot Jupiters, hot Neptunes, and hot Earths have semimajor axes smaller than 0.05 AU and are strongly heated by their parent stars. Tidally locked (or synchronously rotating) planets present the same face to the star at all times (just as the moon presents the same face to the Earth). The permanent day-night sides set up a temperature gradient which may or may not be evened out by atmospheric circulation. The topic of how absorbed stellar energy is circulated in the atmosphere of a tidally locked planet is a complex one. We address it further in Chapter 10.

We limit ourselves to the simplest case: a hot tidally locked planet where the incident stellar energy is absorbed in one layer, instantaneously heats that surface element, and is reemitted. We can solve this problem in analogy with the Lambert sphere scattered light flux ratio developed in section 3.6.1. For a Lambert sphere, the incoming intensity is scattered isotropically. In our ideal case the incident energy heats a surface element and is instantaneously reemitted. Because thermal emission is isotropic, this situation is akin to a Lambert sphere. The normalized phase curve in thermal emission at infrared wavelengths will therefore follow equation [3.58].

Beyond this simplest case, if the exoplanet thermal flux distribution on the planet surface is available from a model atmosphere computation, the thermal IR phase curve can still be calculated from the planet-star flux ratio, where the planet's phase-angle-dependent flux is

$$\mathcal{F}_{\alpha,\mathrm{thermal}}(\nu) = \int_{-\pi/2}^{\pi/2}\int_{-\pi/2}^{\pi/2} I_{\alpha,\mathrm{thermal}}(\theta,\phi,\nu)\cos\phi\cos^2\theta d\theta d\phi. \tag{3.60}$$

Here the subscript "thermal" refers to the thermal infrared component of the flux or intensity.

3.7 SUMMARY

We have taken a first, signficant step towards understanding exoplanetary atmospheres. We described the different definitions of temperature, comparing the definitions by using data on stars and planets. We connected temperature to planetary albedo, via energy balance. A careful derivation of the geometric albedo was compared with a derivation of the Bond albedo, in order to understand the historical definitions and their present relevance to actual exoplanet observations. We derived basic planet-to star flux ratios, using the temperatures and albedos. These planet-to-star flux ratios are useful not just for a basic understanding of planetary radiance properties, but to estimate the feasibility of exoplanet discovery and study with ground- or space-based obserations.

REFERENCES

For further reading

A thorough description of planet albedos is:

- Chapter 9 in Sobolev, G. G., 1975, *Light Scattering in Planetary Atmospheres* (New York: Pergamon Press).

References for this chapter

1. Kurucz Model Atmospheres website http://kurucz.harvard.edu/grids.html
2. Pearl, J. C., and Christensen, P. R. 1997. "Initial Data from the Mars Global Surveyor Thermal Emission Spectrometer Experiment: Observations of the Earth." J. Geophys. Res. 102, 10875–10880.
3. Cox, A. N. 2000. *Allen's Astrophysical Quantities* (4th ed.; New York: Springer).
4. Verbiscer, A., French, R., Showalter, M., and Helfenstein, P. 2007. "Enceladus: Cosmic Graffiti Artist Caught in the Act." Science 315, 815.
5. Sobolev, G. G., 1975. *Light Scattering in Planetary Atmospheres* (New York: Pergamon Press).
6. Rowe, J., et al. 2006. "An Upper Limit on the Albedo of HD 209458b: Direct Imaging Photometry with the MOST Satellite." Astrophys. J. 646, 1241–1251.
7. Conrath, B. J., Hanel, R. A., and Samuelson, R. E. 1989. Origin and Evolution of Planetary and Satellite Atmospheres, 513.

EXERCISES

1. a) Compare the incident energy from stellar radiation to the internal energy for a hot Jupiter exoplanet. Use $T_{\rm eq} = 1600$ K for the planet's equilibrium temperature. Use $T_{\rm eff} = 120$ K for the planet's interior energy (equivalent to Jupiter's interior energy). b) Compare the Earth's received energy from the Sun to its interior energy. The former can be taken as 270 K and the latter is known to be 44×10^{12} W.
2. Derive the radiation constant $\sigma_{\rm R}$ in the Stefan-Boltzmann Law equation [3.6] by integrating the black body over frequency as shown in equation [3.4]. Hint: use the identity

$$\int_0^\infty du \frac{u^3}{e^u - 1} = \frac{\pi^4}{15}. \tag{3.61}$$

3. Derive $T_{\rm eq}$, the equilibrium temperature (equation [3.9]) for a planet on an eccentric orbit.

4. Find five different exoplanets with published $T_{\rm b}(\nu)$. Compute $T_{\rm eq}$ for these planets and compare it to the measured $T_{\rm b}(\nu)$.

5. Show that the fraction of incident energy per unit time scattered at $\alpha = 0$ is $E_{\rm frac} = \frac{1}{\pi} A_{\rm g}$.

6. Show that for a plane Lambert surface, $\Phi_\alpha = \cos\alpha$. Over what angular range is this valid?

7. What is the planet-star flux ratio of a Lambert sphere at $\alpha = 90°$? Use equation [3.58]. Describe conceptually why the flux ratio at $\alpha = 90°$ is less than 1/2 of the flux ratio at $\alpha = 0°$.

8. How does the Lambert sphere phase curve equation [3.58] depend on orbital inclination?

9. Write down an equation for the apparent albedo, introduced in Section 3.4.6. How does the apparent albedo relate to the Bond and spherical albedos?

Chapter Four

Composition of a Planetary Atmosphere

4.1 INTRODUCTION

One of our major goals is to understand the planetary emergent spectrum. Where does the planetary spectrum come from and what does it tell us about the planetary atmosphere? The planetary spectrum is the main observable and therefore our probe into the atmosphere. From a comparison of Earth's, Venus's, and Mars's infrared spectra (Figure 4.1), we can see that the composition of an atmosphere is responsible for the spectral absorption features. What we don't see directly from the spectrum are the abundant gases in a planetary atmosphere that have weak or no absorption or emission features.

In this chapter we discuss the composition of a planetary atmosphere. We start out with Earth's and Jupiter's atmospheric composition. Then we turn to chemistry, used to determine which atoms and molecules will be present in a planetary atmosphere based on the elemental abundance and temperatures and pressures. All solar system planet atmospheres have clouds. We discuss the basic physical conditions for cloud formation as well as the processes that determine cloud particle sizes and vertical extent. Whether a planet retains hydrogen, or whether or not the planet has an atmosphere at all, can be explored with atmospheric escape mechanisms. Last but not least, we touch on atmospheric evolution as the pathway to a planet atmosphere's present chemical composition.

4.2 COMPOSITION OF EARTH'S AND JUPITER'S ATMOSPHERES

The composition of exoplanet atmospheres starts with the elemental abundances. From the elemental abundances, a few dominant molecules will form, depending on the atmospheric temperature (Figure 4.2). In addition to atmospheric gases, clouds will form as some gas species condense into liquid or solid particles, depending on the temperature, pressure, and condensation curve of a given gas. Supporting this overview picture that only a handful of molecules dominate planetary spectra is that in our solar system we see the same dominant molecules in a variety of planetary atmospheres. For example, CH_4 dominates the reflectance spectrum of every giant planet atmosphere. The giant planet infrared spectra also share spectral features from gas species such as NH_3, CH_4, H_2O, and H_2. As a second example, the terrestrial planet (Venus, Earth, Mars) infrared spectra are all dominated by by strong CO_2 absorption (Figure 4.1.)

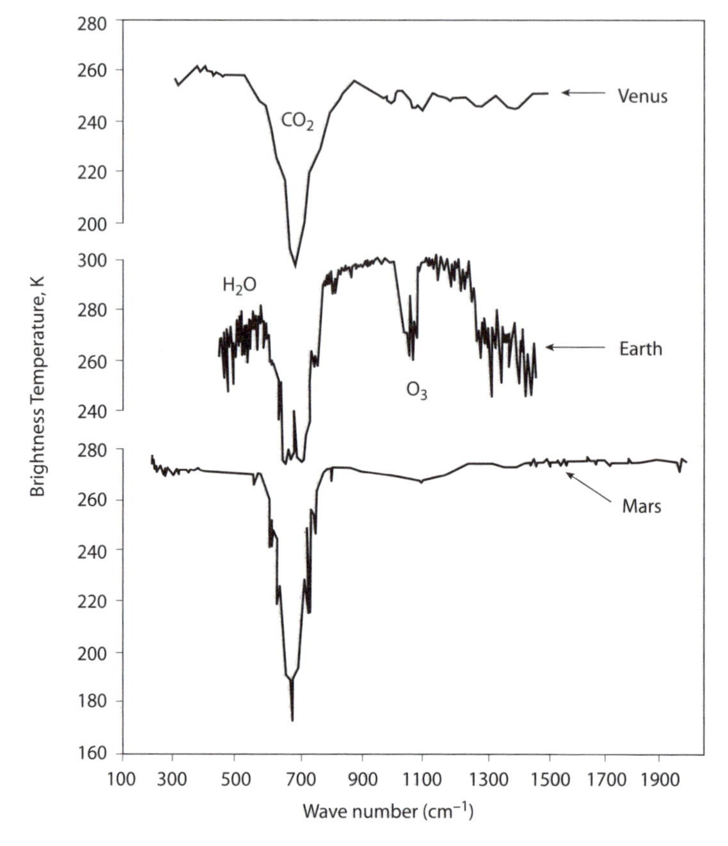

Figure 4.1 Comparison of Venus's, Earth's, and Mars's infrared spectrum. Adapted from [1].

The molecular composition of Earth's lower atmosphere is well known (Figures 4.3 and 4.4 and Table A.2). Molecular nitrogen, N_2, at 78.1% by volume and molecular oxygen, O_2, (at 20.9%) dominate Earth's atmospheric composition. Earth's atmosphere has undergone substantial evolution to end up with N_2 and O_2 as the most abundant atmospheric gases. N_2 is an inert gas that does not react with the surface and is too heavy to escape the Earth's atmosphere and hence has remained as a major component. O_2 is generated by photosynthetic life—plants and photosynthetic bacteria. Because of O_2, Earth's atmospheric composition is far from what we would predict from chemical equilibrium. Earth would have more CO_2 without its surface ocean; most of the CO_2 was dissolved in the ocean to later become sequestered in limestone rocks. Any hydrogen that was likely present in the early Earth would have escaped to space.

N_2 and O_2 are the most abundant gases in Earth's atmosphere, yet the greenhouse gases we always hear about in the news are CO_2, CH_4, and sometimes N_2O. These gases are called greenhouse gases because they are major players in the heating of Earth's atmosphere. Why is there such a discrepancy between the most abun-

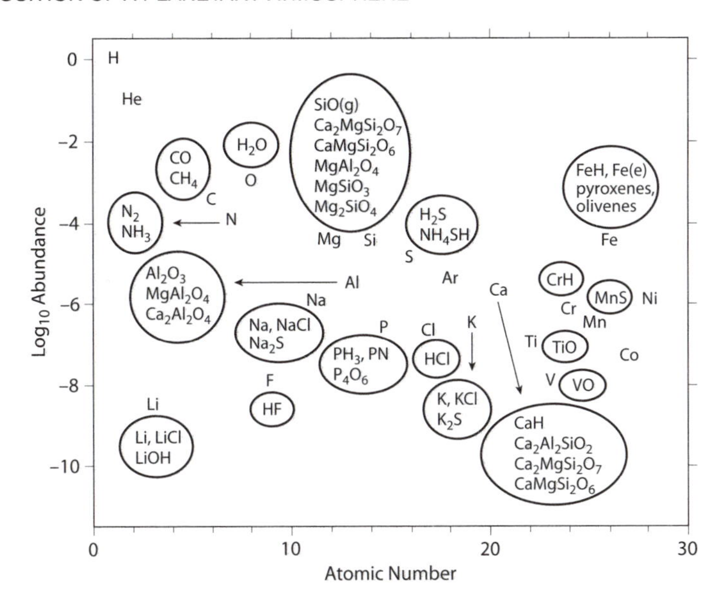

Figure 4.2 Dominant atmospheric molecular and condensate species for individual elements. Unlike giant planet atmospheres, the terrestrial planet atmospheres do not have elements in accordance with solar composition; nevertheless the plot can still be used to see which molecules might dominate. Adapted from [2].

dant gas species in Earth's atmosphere and the gases that are responsible for global warming? The absorbing power depends on the number density times the absorption cross section. Homonuclear molecules such as N_2 and O_2 have very weak absorption cross sections, whereas molecules composed of different atoms such as CO_2 and CH_4 have much stronger absorption cross sections. A very strong absorption cross section means a gas will be a strong absorber, even if it has a low atmospheric abundance.

On Earth, water vapor is the strongest greenhouse gas. Water vapor exists at a few percent or less at Earth's surface and the concentration drops off rapidly above the surface to less than 10 ppm by 40 km altitude. This variable water mixing ratio exists because water vapor is most abundant above the oceans and freezes out in the colder upper troposphere (Figure 4.4). Even though water vapor plays a large role in controlling the atmospheric temperature we do not often hear about water as a greenhouse gas because water vapor is not increasing from human activity.

Ozone, O_3, is another important molecule in Earth's atmosphere. Ozone is present in Earth's stratosphere (and indeed defines the stratosphere; section 9.2); it is formed by photodissocation of O_2.

We digress to define the term atmospheric mixing ratio. The mixing ratio is also called the volume mixing ratio and is defined as the ratio of the number density of a gas to the total number density of the atmosphere. The number density is the number of molecules per unit volume. For example, O_2 in Earth's atmosphere has a mixing ratio of 0.209. CH_4 has a mixing ratio of about 1.7 ppm (parts per million

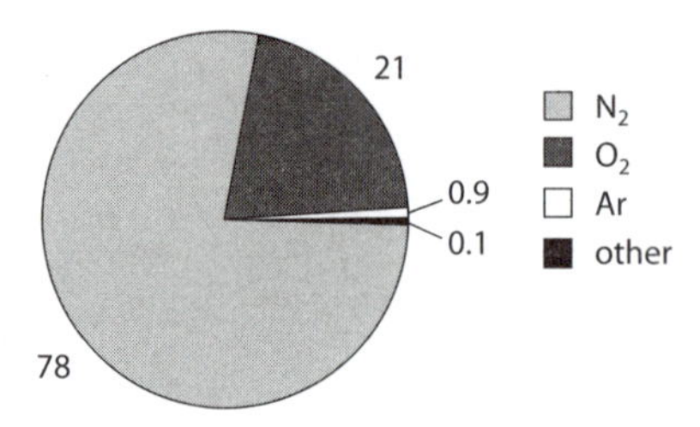

Figure 4.3 Earth's dominant atmospheric composition by approximate percentage volume.

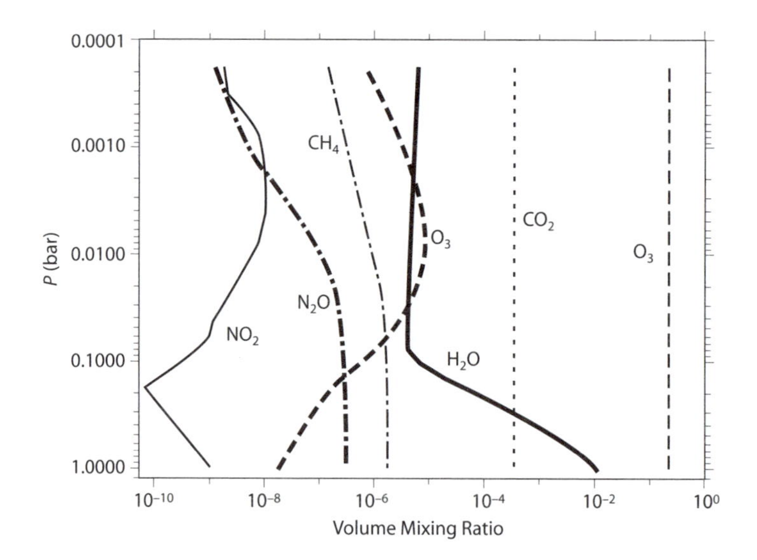

Figure 4.4 Earth's global annual average gas mixing ratios. Adapted from [3].

by volume). The mixing ratio can be altitude dependent for gases that are not well mixed throughout the atmosphere (e.g., Figure 4.4). To complicate matters, on giant planets the mixing ratio is sometimes defined as the ratio of the number density of a gas to the total number density of H_2. We will use X to denote mixing ratio and n to denote number density, sometimes further applying the subscript g to identify the gas in question (e.g., g is CH_4 for methane, NH_3 for ammonia, etc). The mixing ratio is defined as

$$\boxed{X_{\rm g} = \frac{n_{\rm g}}{n_{\rm Total}}.} \tag{4.1}$$

Jupiter's atmospheric composition is very different from Earth's, primarly because Jupiter is hydrogen rich. Jupiter's atmosphere is hydrogen rich because Jupiter formed from the hydrogen-dominated solar nebula and was massive enough to retain all the gases it formed with. Indeed, Jupiter's atmosphere is called primitive because it is the original atmosphere with almost no evolution. Jupiter has an atmospheric composition (in terms of elemental abundances) that basically matches

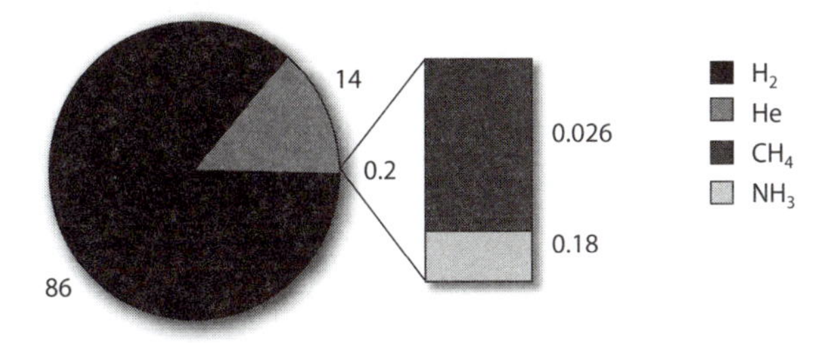

Figure 4.5 Jupiter's dominant atmospheric composition by approximate percentage volume.

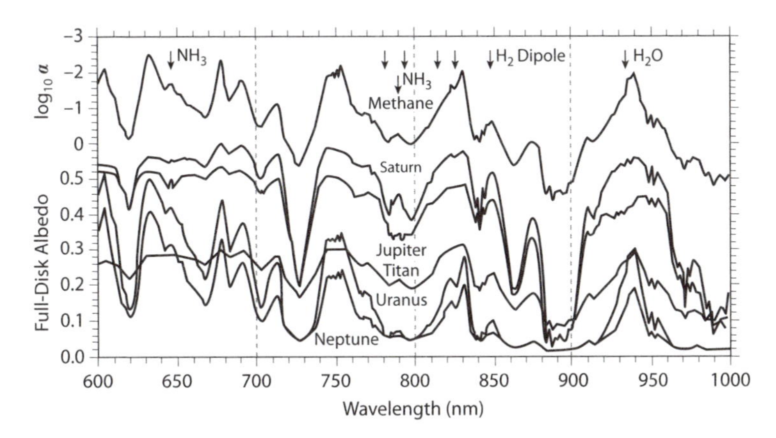

Figure 4.6 Solar system giant planet and Titan visible-wavelength reflection spectra. The methane absorption coefficient is shown in units of km-am^{-1}. Adapted from [4].

that of the Sun and that expected for the solar nebula. Indeed, all of the solar system giant planet atmospheres resemble solar composition (Table A.3) to first order.

On Jupiter, the dominant gases are H_2 and He (see Figure 4.5 and Table A.3). These gases make up 86.2% and 13.6% of Jupiter's atmosphere by volume respectively. H_2 and He are relatively inert species and so (with the exception of collision-induced absorption) do not play a large role as absorbers or emitters and consequently have no deep features in the planetary spectrum. CH_4 has a much lower abundance than H_2 (2.1×10^{-3} as abundant), but is a much stronger absorber, with features dominating the visible-wavelength spectrum (Figure 4.6). The gases in Jupiter's atmosphere are thought to be well mixed, in contrast to some gases in Earth's atmosphere. Clouds of ammonia ices and water ice are present in Jupiter's atmosphere and mostly deplete the cloud-forming gas above the clouds.

The solar system giant planet atmospheres show a deviation from solar abundances. The amount of carbon relative to the solar abundance of carbon increases for increasing planet semimajor axis (Figure 4.7).

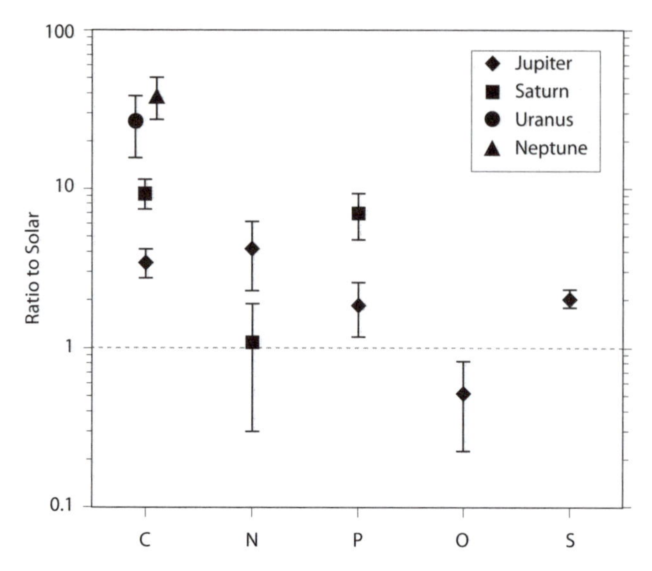

Figure 4.7 Elemental abundances of solar system giant planet atmospheres derived from measurements. Note the logarithmic scale on the y-axis. The carbon abundance increases with increasing semimajor axis of the planets. From [5] and references therein.

4.3 CHEMICAL COMPOSITION

How are we to estimate or determine the atmospheric chemical composition on exoplanets, without the detailed spectra or in situ measurements that are accessible for solar system planets? In the absence of detailed observations, a first step is to adopt an elemental composition and to assume chemical equilibrium.

The number densities of atoms and molecules in a planetary atmosphere are governed by chemistry. Which molecules and atoms are present depends primarily on the elemental abundances and the temperatures and pressures in the atmosphere. The UV radiation from the star also plays a role. Even though this book focuses on physics rather than chemistry, certain concepts in chemistry are fundamental for understanding planetary atmospheres. As an introduction to atmospheric chemistry we will describe some relevant chemical processes. We will proceed in order of increasing complexity from chemical equilibrium to departures from chemical equilibrium to photochemistry.

4.3.1 Chemical Equilibrium

Chemical equilibrium occurs when there is no change in the number densities of atoms and molecules in a closed system. On a microscopic level, chemical reactions continue to occur, but there is no net change. For an atmosphere layer, the closed system is the atoms, molecules, and solids in that layer. Our goal is to know which atoms and molecules exist in an atmospheric layer under chemical equilibrium, given a temperature, pressure, and elemental abundance in that layer. A good rule of thumb is that a closed system wants to reach equilibrium, if it has enough time to do so. We are therefore making a big assumption that the layers in a planet atmosphere have had time to reach chemical equilibrium—an assumption that we will revisit later.

One way to determine the number densities of each species in chemical equilibrium is by minimizing the Gibbs free energy of a closed system. The Gibbs free energy is a thermodynamic function that is extremely useful for predicting the composition of a system at chemical equilibrium.

To fully understand the Gibbs free energy we must delve into thermodynamics. Here we will instead just summarize a few relevant statements about the Gibbs free energy and adopt them. First, the Gibbs free energy does not correspond to a physical property of matter. Second, the Gibbs free energy is somewhat of a misnomer because it is not "energy" but instead an energy potential (and is often referred to as the Gibbs function). Although the Gibbs free energy has the same units as energy, it is not a conserved quantity.

Essentially, the Gibbs free energy arises from a restatement of the first law of thermodynamics. Recall that the first law of thermodynamics, the conservation of energy, can be written in the differential form

$$dU = dQ - dW, \tag{4.2}$$

where dU is the change in internal energy of the system, dQ is the amount of energy added to the system by heating, and dW is the amount of energy lost via work done by the system. We will assume familiarity with the basic concepts $dQ = TdS$, $dW = PdV$ for a reversible process, where the state variables are T (temperature), S (entropy), P (pressure), and V (volume). These definitions can lead us to the following form of the first law of thermodynamics,

$$dU = TdS - PdV. \tag{4.3}$$

This equation still states that the change in energy is related to the internal heating (TdS) and the external work done on the system (PdV).

The problem in trying to use the first law of thermodynamics for chemical equilibrium in a planetary atmosphere is that it is not possible to measure the change in entropy dS and the change in volume dV. We would prefer a formulation that uses the change in temperature dT and the change in pressure dP where we can relate the change in pressure to the partial pressure of a given atomic or molecular species. Even though we cannot directly measure T, dT, P, and dP in exoplanetary atmospheres, they are quantities that we can infer directly from spectral measurements.

The Gibbs free energy, then, arises from the need to express the first law of thermodynamics that uses variable dT and dP, quantities more relevant to planetary atmospheres than dS and dV. The Gibbs free energy is defined as

$$G = U + PV - TS. \tag{4.4}$$

The differential form is

$$\boxed{dG = VdP - SdT,} \tag{4.5}$$

where we have used equation [4.3] to subsitute for dU.

We now want to understand why chemical equilibrium means a minimum of the Gibbs free energy. At equilibrium the temperature and pressure are constant ($dP = dT = 0$). Given the formulation for the Gibbs free energy as a variant of the first law of thermodynamics (equation [4.5]) we see that at equilibrium $dG = 0$. We also invoke the second law of thermodynamics that the entropy of a closed system tends to increase or stay the same, $dS \geq dQ/T$. We can then write equation [4.3] as

$$TdS \geq dU + PdV \tag{4.6}$$

and equation [4.5] as

$$dG \leq VdP - SdT. \tag{4.7}$$

We can see that for a chemical change at constant temperature and pressure

$$dG \leq 0. \tag{4.8}$$

In other words, the Gibbs free energy of a system at constant temperature and pressure will tend to decrease. As long as $dG < 0$, the system will drive G to smaller and smaller values. At some point G will reach a minimum, and there dG becomes zero.

Let us now take the one-component example of liquid and solid water as a function of temperature (Figure 4.8). In thermodynamic equilibrium, and knowing the Gibbs free energy as a function of temperature, we can tell whether water is in the liquid or solid phase. For example, at temperatures where the Gibbs free energy is lower for liquid water than for water ice, water will be in liquid form. (See below for the pressure dependence of the Gibbs free energy.)

For a gas with two or more components, an additional term must be added to the first law of thermodynamics to account for the changing number of different particles (such as atoms and molecules), $\sum_i \mu_i dN_i$. Here μ_i is the chemical potential: the amount of energy added to the system for particle i (and not to be confused with angle cosines μ used elsewhere in this book). Here N_i refers to the number of particles of a given species i. The first law of thermodynamics is then

$$dU = TdS - PdV + \sum_i \mu_i dN_i. \tag{4.9}$$

A Gibbs free energy chemical equilibrium calculation for a number of species (hundreds or even more) involves finding the minimum energy for all molecules and atoms of interest. The species can possibly include ions and solids as well. The general idea is take tabulated temperature-dependent Gibbs free energies g^0 of individual gas species i, and scale them with pressure to describe the Gibbs free energy (G) of a system,

$$G = \sum_i n_i \left(g_i^0(T) + RT \ln P_i \right), \tag{4.10}$$

where n_i is the number of moles. The term $RT \ln P_i$ describes the pressure dependence with R the universal gas constant and P_i the partial pressure of species i, and comes from the ideal gas law.

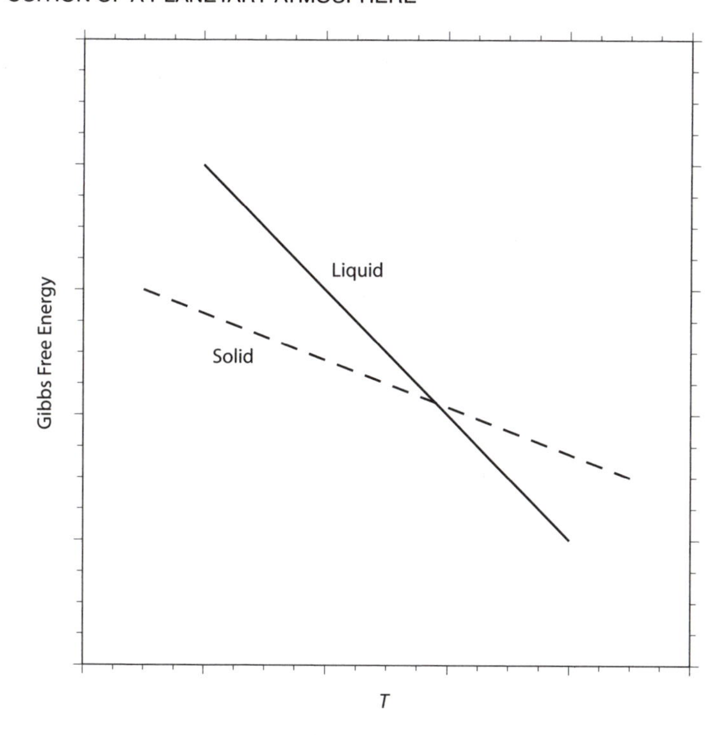

Figure 4.8 Schematic diagram of the Gibbs free energy of liquid and solid water. At low temperatures (on the left side of the diagram) solid water (ice) dominates because it has a lower Gibbs free energy. At higher temperatures (right side of the diagram), water is in liquid form. The Gibbs free energies of liquid and solid water cross at 273 K.

The dimensionless Gibbs free energy equation to be minimized is

$$\boxed{G/RT = \sum_i n_i \left[(g_i^0(T)/RT) + \ln P + \ln(n_i/N)\right],} \tag{4.11}$$

subject to the number density constraint equation

$$\sum_i a_{ij} n_i = b_j. \tag{4.12}$$

Here N is the total number of moles in the system, and we have used $P_i/P = n_i/N$. Furthermore, a_{ij} is the number of atoms of type j in a molecule considered for n_i, and b_j is the total number of moles of element j. The solution of the system is a non-negative set of values n_i such that equation [4.11] is minimized. To include gas ions, a charge constraint equation can be added.

Chemical equilibrium helps us to understand the atmospheric composition of giant planets. Jupiter and Saturn are dominated by H_2, and CH_4 is the dominant form of carbon. For solar abundances hot Jupiters are expected to have CO as the dominant form of carbon and N_2 as the dominant form of nitrogen.

To include solid particles, the Gibbs free energies of solids are added to equation [4.11]. These do not have the pressure term. After cloud formation is considered, the Gibbs free energy is minimized with a system depleted of elements that go into the cloud particles.

4.3.2 Departures from Chemical Equilibrium

Under what conditions is chemical equilibrium reached? Generally speaking, if there is enough time for reactions to proceed fully in a closed system, then chemical equilibrium will be reached. Reactions proceed faster at higher temperatures and pressures and so chemical equilibrium is most likely to be reached in the deep, hotter parts of planetary atmospheres. In contrast, chemical disequilibrium is more likely to prevail in the upper, usually cooler, lower-density regions of planetary atmospheres where nonequilibrium processes have shorter time scales than equilibrium processes. Nonequilibrium chemistry is ubiquitous in solar system planet atmospheres.

Brown dwarfs provide two good examples of chemical disequilibrium [6]. Brown dwarfs are failed stars whose interiors never reached high enough temperatures and pressures to sustain stable hydrogen fusion. Brown dwarfs did, however, burn some amount of deuterium at some stage in the past, leading to residual interior energy that heats the brown dwarf as it slowly leaks out through the brown dwarf interior and atmosphere. With effective temperatures ranging from 600 to 2000 K, brown dwarfs are somewhat useful analogs to hot Jupiters because brown dwarfs have atmospheres in the same temperature range. However, the atmospheric structure is different because brown dwarf atmospheres are heated from the interior whereas the hot Jupiter atmospheres are heated from above the atmosphere, externally from the parent stars.

The first brown dwarf chemical disequilibrium example involves nitrogen chemistry. In a hydrogen-rich atmosphere at the temperatures of some brown dwarfs, the dominant form of nitrogen is ammonia, NH_3, for an atmosphere in chemical equilibrium. Observations of brown dwarfs, however, show an underabundance of NH_3, at the level of one order of magnitude. In other words, N_2 and NH_3 are far out of chemical equilibrium in brown dwarf atmospheres. The reaction describing the N_2 to NH_3 conversion is

$$N_2 + 3H_2 \leftrightarrow 2NH_3. \tag{4.13}$$

The reason NH_3 and N_2 are out of chemical equilibrium is related to the N_2 triple bond. The triple bond has a large binding energy. The above reaction, equation [4.13], proceeds more slowly to the right than to the left, because the N_2 triple bond is difficult to break apart. The explanation for the overabundance of N_2 in brown dwarfs is that vertical mixing dredges up N_2 from deep in the atmosphere where chemical equilibrium favors N_2. In the cooler, upper atmosphere, N_2 is converted very slowly to NH_3. The rate for dredging up N_2 is much faster than the reaction rate to convert N_2 to NH_3, so that chemical equilibrium is never reached in the observable atmosphere, and NH_3 is under abundant.

A second brown dwarf chemical disequilibrium example involves carbon chemistry. For "T type" and cooler brown dwarfs, carbon monoxide is the dominant form of carbon at the higher temperatures deep in a brown dwarf atmosphere whereas methane is the dominant form of carbon in the cooler upper atmospheres. Yet, similar to the nitrogen chemistry example, CO is overabundant compared to equilibrium chemistry predictions. As in the situation caused by the N_2 triple bond, the CO double bond has a high binding energy, making the CO to CH_4 conversion reaction very slow. Under the same scenario of vertical dredging from the hot interior that is faster than the CO to CH_4 conversion, chemical equilibrium is never reached. The relevant reaction is

$$\mathrm{CO} + 3\mathrm{H}_2 \leftrightarrow \mathrm{CH}_4 + \mathrm{H}_2\mathrm{O}, \tag{4.14}$$

and it proceeds more slowly to the right than to the left because of the strong CO double bond. Jupiter shares this carbon disequilibrium example, because CO should not exist at the cold temperatures in Jupiter's atmosphere. The actual presence of CO in Jupiter's cold atmosphere attests to deep vertical atmospheric mixing.

To describe an atmosphere with parts that deviate from chemical equilibrium we may consider an atmosphere divided into two vertical regions: a deep lower region in chemical equilibrium, and an upper, cooler region that departs from chemical equilibrium. The lower region is defined where the timescale for chemical equilibrium $\tau_{\rm chem}$ is shorter than the vertical mixing timescale $\tau_{\rm mix}$ (here the timescale τ should not be confused with the optical depth τ used in other chapters). In other words, even if new chemical species are brought to this lower region (from the interior or the upper atmosphere), the chemical reaction rates proceed fast enough for chemical equilibrium to be reached. In the upper region, the opposite occurs, $\tau_{\rm chem} \gg \tau_{\rm mix}$. The regions are divided at the level where $\tau_{\rm chem} = \tau_{\rm mix}$, the so-called quench level.

Let us use the N_2-NH_3 example described above for brown dwarf atmospheres, to further explore the description of an atmosphere with parts that deviate from chemical equilibrium. In the lower atmosphere region we can assume that N_2 and NH_3 take on their chemical equilibrium values. One or the other may dominate (depending on the temperature and pressure), and there is likely to be some of each present. In the upper region where $\tau_{\rm chem} > \tau_{\rm mix}$, one can assume that the N_2 and NH_3 values take on the same ratios as at the level where $\tau_{\rm chem} = \tau_{\rm mix}$. Each different molecular species has its own level where $\tau_{\rm mix} = \tau_{\rm chem}$. See Figure 4.9 for an example.

To describe the term $\tau_{\rm mix}$, we must use some terminology that is defined later in this or other chapters. $\tau_{\rm mix}$ is defined differently depending on whether radiation or convection is the dominant energy transport mechanism at the altitude in question. In the radiative zone, $\tau_{\rm mix} \sim H^2/\kappa_{\rm E}$, where H is the pressure scale height in length units of m (equation [9.11]) and $\kappa_{\rm E}$ is the eddy diffusion coeffecient in units of $\mathrm{m}^2\ \mathrm{s}^{-1}$ (Section 4.3.3). $\kappa_{\rm E}$ cannot be experimentally determined and is unknown for exoplanet atmospheres; $\kappa_{\rm E}$ is usually taken as a free parameter in models. In the convective zone, $\tau_{\rm mix} \sim H/v_{\rm c}$, where $v_{\rm c}$ is the convective velocity (see Section 9.8.3). Note that mixing in the convective zone is always much faster than in the radiative zone, because convection moves material faster than eddy diffusion.

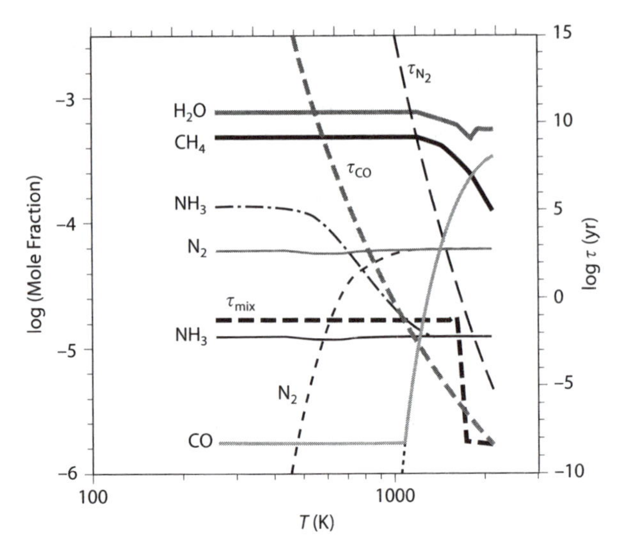

Figure 4.9 Chemical profile for a model atmosphere of the brown dwarf Gl 570D. Mole fractions of molecules are shown in equilibrium (dashed, dotted, or dash-dot curves) and out of equilibrium (solid curves). The mole fractions are shown as a function of temperature. The brown dwarf temperature increases deeper into the atmosphere, and so the pressure increases from the top of the atmosphere (left) toward the bottom (right). The thick dashed lines show the mixing timescale ($\tau_{\rm mix}$) and the timescale for the destruction of CO ($\tau_{\rm CO}$) and N_2 (τ_{N_2}). Adapted from [6].

4.3.3 Photochemistry

For many exoplanets the star's UV radiation will play a major role in determining the chemical composition of the planetary atmosphere. The star's UV radiation can photodissociate molecules in the upper layers of the planetary atmospheres. The atoms resulting from the split apart molecules might escape to space or may recombine to form a different molecule. In this way photochemistry is essential to an understanding of the planetary atmosphere composition.

Photochemistry is most signficant for relatively low-mass planets. On massive planets (such as Jupiter and Saturn) photochemistry does not radically alter the atmospheric composition for two reasons. First, the large planet mass makes atmospheric escape negligible and, second, deep in the atmosphere chemical reaction rates are fast. Fast reaction rates in a well-mixed atmosphere enable reformation of the molecules split apart by photochemistry. In contrast, on a low-mass planet such as Earth, the light gases H and He escape from the atmosphere into space. Earth's atmosphere ends at the solid surface, and temperatures and pressures never get high enough in the deepest part of the atmosphere for the return path to form the molecules that are split up by photochemistry.

The overview for photochemistry is that we write down 1D rate equations for individual atoms or molecules whereby the rate of change is equal to the sources and sinks for that molecule,

$$\boxed{\frac{\partial n_i(z)}{\partial t} + \frac{\partial \phi_i(z)}{\partial z} = P_i(z) - n_i(z)L_i(z).} \tag{4.15}$$

Here $n_i(z)$ is the number density of a given species i at altitude z, and the subscript i is used to imply that each species i has its own photochemical equation. A mixing ratio can be used instead of number density. The sources of $n_i(z)$ are collected in the production rate term $P_i(z)$ and the sinks of $n_i(z)$ are collected in the loss rate term $L_i(z)$. Production and loss rates can come from photochemical or chemical reactions. For terrestrial planets production rates also include surface fluxes generated by either geochemical processes or life and loss rates include escape to space.

There is never just one single equation in photochemistry, because the sources and sinks themselves involve other chemical species. This means that the equations for each species i can be coupled through the production and loss rate terms. The number densities of atoms and molecules in each layer z are connected by the vertical movement or "flux" of that particle, $\phi_i(z)$. We will drop the flux term for our initial discussion of photochemistry and revisit it later.

Let us take a canonical concrete example for a starting point for understanding photochemistry. We will use a generic atom A, diatomic molecule A_2, and a third atom or molecule we will conventionally call M. Taken together with a simple, hypothetical network of photochemical reactions in a homogeneous atmosphere layer,

$$A_2 + h\nu \xrightarrow{J} A + A, \tag{4.16}$$

$$A + A + M \xrightarrow{K} A_2 + M. \tag{4.17}$$

Here J is the photodissociation rate and K is the chemical reaction rate. We then write the corresponding hypothetical rate equations

$$\frac{\partial A}{\partial t} = +J[A_2] - K[A][A] = 0, \tag{4.18}$$

$$\frac{\partial A_2}{\partial t} = -J[A_2] + K[A][A] = 0, \tag{4.19}$$

where the square brackets denote concentrations and we have adopted photochemical equilibrum to assume $\frac{\partial A}{\partial t} = \frac{\partial A_2}{\partial t} = 0$. We also have a number density conservation equation where we have a fixed, total amount of the element A, which we call A_{total},

$$A + 2A_2 = A_{\text{total}}. \tag{4.20}$$

We want to solve for the unknown number densities, A and A_2, and we require two equations. We use the number conservation equation, and only one of equations [4.18] and [4.19] because they are redundant with each other. We rewrite equation [4.18], to have the following equations to solve:

$$[A] = \sqrt{\frac{J[A_2]}{K}}, \tag{4.21}$$

$$A + 2A_2 = A_{\text{total}}. \tag{4.22}$$

Here A_{total} is an input to the model.

The quantity J, in units of s^{-1}, is the photodissociation rate

$$J(z) = \int_0^\infty \sigma(\nu) F_{act}(\nu) d\nu, \tag{4.23}$$

where z is altitude in the vertical direction, ν is the frequency and $\sigma(\nu)$ is the absorption cross section. Here $F_{act}(\nu)$ is the spectral actinic flux, the flux incident on the molecule from all directions, in units of photons $m^{-2}\ s^{-1}\ m^{-1}$. The spectral actinic flux can conceptually be related to the exponentially attenuated incident stellar flux. For our purposes, it is relevant that the cross sections are strong at extreme ultraviolet (EUV, ~10–100 nm) or ultraviolet (UV, ~100–320 nm) wavelengths. It is also relevant that the stellar flux is usually absorbed by these strong cross sections high in the atmosphere, well above millibar levels, and sometimes above microbar levels. Therefore, photochemistry for solar system planets is limited to the upper atmosphere.

We now turn to a more realistic system of equations, that of ozone abundance in a layer in Earth's atmosphere above about 30 km (following [7]). The network of relevant equations is

$$\mathrm{O} + \mathrm{O_2} + \mathrm{M} \xrightarrow{K_{12}} \mathrm{O_3} + \mathrm{M}, \tag{4.24}$$

$$\mathrm{O_2} + h\nu(\lambda < 242\ \mathrm{nm}) \xrightarrow{J_2} \mathrm{O} + \mathrm{O}, \tag{4.25}$$

$$\mathrm{O_3} + h\nu(\lambda < 1100\ \mathrm{nm}) \xrightarrow{J_3} \mathrm{O} + \mathrm{O_2}, \tag{4.26}$$

$$\mathrm{O} + \mathrm{O_3} \xrightarrow{K_{13}} 2\mathrm{O_2}, \tag{4.27}$$

where the first two equations are ozone formation pathways and the last two ozone destruction pathways. M is any third atom or molecule. We can write the following rate equations for the number densities of O and O_3:

$$\frac{\partial[\mathrm{O}]}{\partial t} = -[\mathrm{O}][\mathrm{O_2}][\mathrm{M}]K_{12} + 2[\mathrm{O_2}]J_2 + [\mathrm{O_3}]J_3 - [\mathrm{O}][\mathrm{O_3}]K_{13}, \tag{4.28}$$

$$\frac{\partial[\mathrm{O_3}]}{\partial t} = [\mathrm{O}][\mathrm{O_2}][\mathrm{M}]K_{12} - [\mathrm{O_3}]J_3 + [\mathrm{O}][\mathrm{O_3}]K_{13}. \tag{4.29}$$

We add the above two equations together to find

$$\frac{\partial[\mathrm{O}]}{\partial t} + \frac{\partial[\mathrm{O_3}]}{\partial t} = 2[\mathrm{O_2}]J_2 - 2[\mathrm{O}][\mathrm{O_3}]K_{13}. \tag{4.30}$$

We can rewrite equation [4.29] with the knowledge that below 75 km $[\mathrm{O}][\mathrm{O_3}]K_{13} \ll [\mathrm{O}][\mathrm{O_2}][\mathrm{M}]K_{12}$,

$$\frac{\partial[\mathrm{O_3}]}{\partial t} \simeq [\mathrm{O}][\mathrm{O_2}][\mathrm{M}]K_{12} - [\mathrm{O_3}]J_3. \tag{4.31}$$

Taking the steady state of photochemical equilibrium $\frac{\partial[\mathrm{O}]}{\partial t} = \frac{\partial[\mathrm{O_3}]}{\partial t} = 0$, we can combine the above two equations to find

$$[\mathrm{O_3}] \simeq \sqrt{\frac{J_2 K_{12}}{J_3 K_{13}}[\mathrm{M}][\mathrm{O_2}]}. \tag{4.32}$$

We can use the above equation to see that the amount of ozone is low when the ozone destruction rate J_3 is high. This happens in relatively high layers where there is lots of UV radiation from the Sun. We can similarly see that ozone concentration is high in layers where the O_2 destruction rate J_2 is high; this step starts the ozone formation process. This happens slightly lower in the atmosphere where the slightly longer-wavelength energy penetrates.

For planetary atmospheres we are almost always unable to reduce a network of photochemical equations to a single analytic equation. This is especially the case for exoplanet atmospheres because we are not able to measure the concentrations of any individual species, and the species are connected to each other through many equations. A full photochemical network of equations typically includes one hundred or more reaction rates. Some photochemical reactions generate aerosols (such as the destruction of SO_2 to form sulfate aerosols).

Until now we have considered photochemical reactions in a single vertical layer in a planet atmosphere. Atmosphere layers are connected vertically, however, through mixing in the atmosphere. Mixing in the region where photochemistry dominates is not usually convective mixing. Convection occurs much deeper in the atmosphere where the star's UV radiation does not penetrate to cause photochemistry. Molecular diffusion is also not a signficant mixing mechanism in the layers where photochemistry is relevant. Molecular diffusion is important in layers high in the atmosphere (the "thermosphere" on Earth), a layer that for exoplanets is not readily detectable with observations.

Mixing in the atmosphere layers where photochemistry is occurring is caused by a variety of microscopic physical processes. The eddy diffusion coefficient κ_{E} is used as a macroscopic coefficient to parameterize the diffusion processes. Because it is difficult to measure κ_{E} experimentally, modelers use a range of values. In Jupiter's and Saturn's atmospheres $\kappa_{\mathrm{E}} = 10^3$ to 10^5 m^2 s^{-1}, as determined from model interpretation of the observed heat fluxes. For exoplanets and brown dwarfs that lack many concrete observations, the range can span several orders of magnitude.

The number density flux term $\phi_i(z)$ in equation [4.15] is

$$\phi_i(z) = n_i(z)v_i = \kappa_{\mathrm{E}}\frac{\partial n_i}{\partial z}, \tag{4.33}$$

where v_i is the velocity in the vertical direction, so that the term in the photochemistry equation [4.15] is

$$\frac{\partial \phi_i(z)}{\partial z} = -\frac{\partial}{\partial z}\left[\kappa_{\mathrm{E}}\frac{\partial n_i}{\partial z}\right]. \tag{4.34}$$

Additional terms can be added to the number density flux term when relevant, such as molecular diffusion.

Photochemistry plays an important role in many different kinds of planet atmospheres. On early Venus, photochemistry aided the runaway loss of liquid water oceans, by photodissociating water vapor in the atmosphere with the subsequent escape of H. On Jupiter and Saturn, photochemistry creates hazes which significantly mute the blue part of the spectrum (Figure 4.6). Specifically, UV radiation photodissociates some CH_4, and the C and H atoms recombine to form higher hydrocarbons which form hazes. On hot Jupiters, detection of CO_2 is thought to be caused by photochemistry, because chemical equilibrium models predict that CO is the dominant form of carbon, with smaller amounts of CH_4 and very small amounts of CO_2.

4.4 BASIC CLOUD PHYSICS

4.4.1 Cloud Overview

Clouds are present on every solar system planet with an atmosphere. Clouds are so important because of their contribution to atmospheric energy balance. We saw in Chapter 3 that a planet's albedo controls how much stellar energy a planet atmosphere can absorb. Some clouds are highly reflective, contributing to a high albedo, such as Earth's water clouds and Venus's hydrosulfuric acid clouds. Other types of clouds, such as hazes on Jupiter and Saturn, are strongly absorbing at short wavelengths. Clouds are so interesting because of the color and variability they can add to a planet atmosphere. The color of Jupiter's bands and the photometric variability of Earth as viewed from afar are both caused by clouds. Clouds are also fascinating because cloud formation is intricately linked to atmospheric dynamics and temperature and in principle should provide clues to atmospheric behavior.

We define clouds as a mass of liquid or solid particles suspended in a planet atmosphere. If a cloud is present, some of the atmospheric gas may have condensed into the liquid or solid particles. Alternatively, clouds may be photochemically produced, that is, generated when a photochemically produced gas condenses. The terminology of clouds, aerosols, and hazes can be confusing because the terms are often used interchangeably. Aerosols are defined as particles suspended in the atmosphere. Aerosols can include any size and kind of particle, including smoke, pollutants, bacteria, and even cloud particles. Even though "aerosol" technically includes a particle of any size and composition, aerosol is almost always used to refer to very small particles that do not precipitate out of the atmosphere (and not to the larger particles that can be found in clouds). Haze is defined as very small particles suspended in the atmosphere that are opaque, diminish visibility, and contribute to color. Hazes are almost always photochemically produced.

The diversity of cloud types on different planets is enormous—the range of cloud composition, vertical and horizontal extent, and particle size is huge. Earth's clouds are made of liquid water droplets or frozen water crystals ranging from 10 to 100 μm in size. Titan has clouds likely composed of liquid methane. We think the hot Jupiter exoplanets may have clouds composed of liquid iron droplets, solid silicate particles, or other kinds of material that condenses at high temperatures. A single planet can have different cloud layers if its atmosphere spans a temperature range that includes more than one condensible gas. Jupiter, as illustrated in Figure 4.10, has several different cloud layers. Figure 4.10 also shows cloud layers in giant planets that are progressively warmer than Jupiter (but with the same composition). It is interesting that hot Jupiters can have silicate or iron clouds, whereas the same kinds of clouds are almost certainly present deep in Jupiter's atmosphere where the temperatures are similar.

Clouds pose a conundrum for exoplanets. On the one hand, the existence of

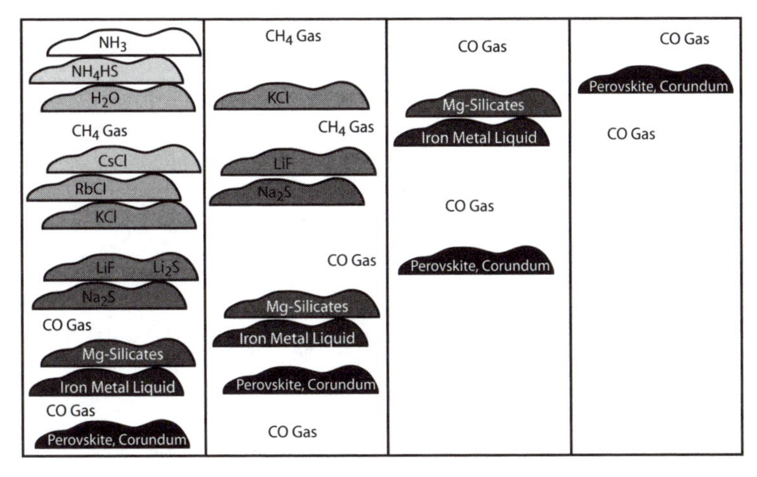

Figure 4.10 Illustration of cloud layers in giant planet atmospheres. The top of the figure represents the top of the planet atmosphere. The left panel shows clouds expected to be present in Jupiter's atmosphere. The second panel from the left is for a warmer giant planet that is too warm for water clouds. The third panel from the left and the right panel are for increasingly hotter giant planets, hypothetical hot Jupiters. The planets warmer than Jupiter can still have clouds, just of very different composition. Notice how the clouds are confined to layers, and that clouds of the same composition are present at higher altitudes in warmer planets. Adapted from [9].

clouds has the potential to tell us a good deal about a planet atmosphere: the temperature and possibly even the planet's rotation rate for relatively fixed large-scale cloud patterns. On the other hand, interpretation of exoplanet atmosphere data is made so much more complex by clouds. First, clouds can block the atmosphere beneath them, weakening the emergent spectral features. Second, clouds are difficult to model, because prediction of cloud properties such as cloud vertical height, fraction of gas condensed, and particle size distribution is complicated. Last but not least, radiative transfer through clouds can be challenging.

Exoplanet atmosphere models have seen a variety of different cloud treatments. These range from cloud formation models to estimate the cloud vertical extent and particle size, to treatments of the cloud vertical extent, density, and particle size of a fixed distribution. Most models to date adopt the possibly unrealistic assumption of complete cloud cover for homogeneous clouds; a few attempts beyond this have been made, for example, clouds as a fractal pattern. Evidence for haze on the exoplanet HD 189733 was obtained with the *Hubble Space Telescope* [8].

4.4.2 Cloud Altitude

Condensation clouds form at an altitude in a planetary atmosphere where a gas can condense into a liquid or solid. Let us take the example of water clouds on Earth. First we must look ahead (to Section 9.2) to know that above Earth's surface, the atmospheric temperature drops with increasing altitude above the Earth's surface. When warm air laden with water vapor rises, it expands and cools. If the water vapor cools to a temperature where water vapor can condense, a condensation cloud will form. Water vapor can condense where the partial pressure of water vapor exceeds the saturation vapor pressure. Here "water vapor pressure" is the partial pressure of water gas in the atmosphere and the "saturation vapor pressure" is the water vapor partial pressure when the air is fully saturated.

For any gas species that might condense,

$$P_{\rm g}(T) = X_{\rm g} P(T) \geq P_{\rm Vg}(T), \tag{4.35}$$

where $P(T)$ is the total gas pressure, $P_{\rm g}$ is the gas partial pressure, $X_{\rm g}$ is the mixing ratio, and $P_{\rm Vg}$ is the saturation vapor pressure. The subscript g refers to the gas under consideration.

Cloud base altitudes for different solar system planets are shown in Figure 4.11 (from [10], which we also adapt in the following discussion). The cloud base is at the intersection of the saturation vapor pressure curve and the planet's temperature-pressure profile. The saturation vapor pressure curve is the equilibrium curve between the gas and liquid or solid.

The saturation vapor pressure as a function of temperature is known as the Clausius-Clapeyron curve,

$$\frac{dP_V}{dT} = \frac{L}{T(V_2 - V_1)}, \tag{4.36}$$

where $V = 1/\rho$ is the specific volume, subscript 1 refers to the liquid or solid phase and subscript 2 refers to the vapor phase. L is the specific latent heat of the phase transition in units J kg^{-1}. The Clausius-Clapeyron curve can be derived from thermodynamics, specifically from the first and second laws of thermodynamics. We can simplify the above equation by considering that the vapor is an ideal gas (appropriate for planetary atmospheres), and that the specific volume of the solid or liquid phase is small compared to that of the vapor phase, that is, $(V_2 - V_1) \approx V_1 = R_V T/P$, where R_V is the specific gas constant for the vapor in question [10]. The approximate form of the Clausius-Clapeyron equation follows from equation [4.36],

$$\frac{dP_V}{dT} = \frac{L P_V}{R_V T^2}. \tag{4.37}$$

To proceed we need an equation for L, the specific latent heat of the phase transition. We can integrate the expression for L,

$$\left(\frac{\partial L}{\partial T}\right)_P = \Delta c_p, \tag{4.38}$$

where Δc_p is the change of the specific heat capacity at constant pressure (in units of J kg^{-1} K^{-1}) between the vapor and solid or liquid phase. The specific heat capacity can be expanded by $c_p(T) = \alpha + \beta T$, yielding after integration

$$L = L_0 + \Delta\alpha T + \frac{1}{2}\Delta\beta T^2. \tag{4.39}$$

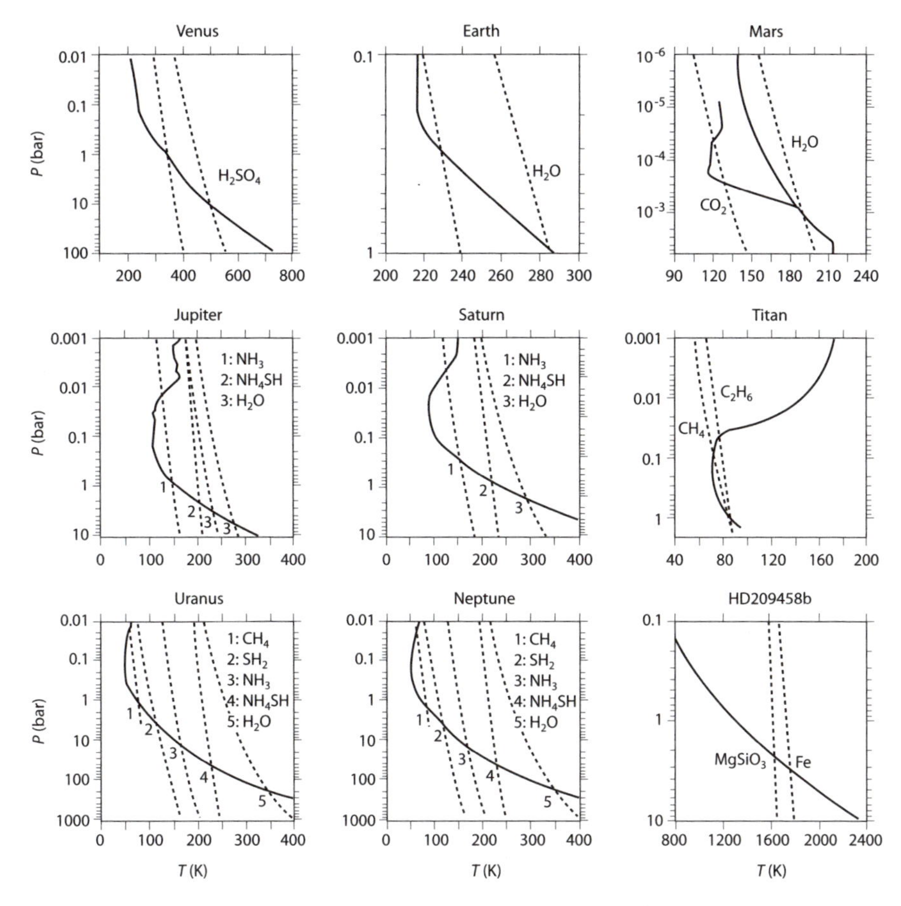

Figure 4.11 Saturation vapor pressure curves (dashed lines) compared to planet vertical temperature profiles. The saturation vapor pressure curves are generated from the Clausius-Clapeyron relationship. The intersection of a saturation vapor pressure curve and a planet pressure-temperature profile marks the cloud base. In some cases two saturation curves are given, corresponding to two limiting abundance cases. Two pressure-temperature profiles are given for Mars because of its seasonal variablity; the right curve is a yearly average and the left curve is a cold profile to illustrate CO_2 condensation. Adapted from [10].

Here L_0 is an integration constant. $\Delta\alpha$ and $\Delta\beta$ are the changes of the constants α and β between the gas and liquid or solid phase, and are determined empirically. To reach a useful form of the Clausius-Clapeyron equation we can subsitute the temperature-dependent expression for L (equation [4.39]) into equation [4.37] to find a general form for the saturation vapor pressure curve,

$$\boxed{\ln(P_V) = \ln(C) + \frac{1}{R_V}\left[-\frac{L_0}{T} + \Delta\alpha \ln T + \frac{\Delta\beta}{2}T\right].} \quad (4.40)$$

Here C is in pressure units and is an integration constant, and we have dropped higher-order temperature terms. Values for the constants are provided in Table A.6.

The beauty of considering cloud formation by the intersection of a saturation vapor pressure condensation curve with an atmospheric temperature-pressure profile is the understanding that a huge variety of clouds can occur on exoplanets, depending on the range of atmospheric pressure and composition. For example, hot Jupiters are far too hot for water clouds and instead are anticipated to have silicate and iron clouds. The hotter brown dwarfs are also known to have clouds of high-temperature condensates.

The condensation curve framework assumes that the vapor is always at 100% humidity and that excess vapor has condensed into a liquid or solid. In reality, cloud formation requires condensation nuclei. In Earth's atmosphere the condensation nuclei are particles of dust or soot. In exoplanet atmospheres, the populations of nuclei on which cloud droplets form are not known.

4.4.3 Cloud Vertical Extent and Density

Condensation curves tell us where a cloud can form, but provide no information about particle sizes and cloud vertical extent. The cloud vertical extent can be estimated by using the Clausius-Clapeyron relationship in equation [4.37]. To estimate the cloud vertical extent we can integrate the equation from a cloud base temperature $T_{\rm c}$ to a temperature T higher in the atmosphere. Approximating L as constant, we find

$$P_V(T) = P_V(T_{\rm c}) \exp\left[-\frac{L(T_{\rm c} - T)}{R_V T T_{\rm c}}\right]. \tag{4.41}$$

We can make some further approximations, first taking $T \sim T_{\rm c}$ to give $TT_{\rm c} \sim T_{\rm c}^2$. Second, we must look ahead to the temperature profile in a convective atmosphere (equation [9.24]),

$$\frac{dT}{dz} = -\frac{g}{c_p}. \tag{4.42}$$

We can therefore write

$$T_{\rm c} - T = \frac{g}{c_p}(z - z_{\rm c}), \tag{4.43}$$

to find

$$P_V(T) = P_V(T_{\rm c}) {\rm e}^{-(z-z_{\rm c})/H_{\rm c}}, \tag{4.44}$$

where we have defined the cloud vertical scale height as

$$H_{\rm c} = \frac{R_V T_{\rm c}^2 c_p}{gL}. \tag{4.45}$$

The cloud scale height is the distance over which the gas saturation vapor pressure (and by extension density) drops off by a factor of e. Again looking forward, we will find the atmospheric pressure scale height (derived from hydrostatic equilibrium; equation [9.11]),

$$H = \frac{kT}{\mu_{\rm m} m_{\rm H} g} = \frac{R_{\rm s} T}{g}, \tag{4.46}$$

where k is Boltzmann's constant, μ_{m} is the mean molecular weight, m_{H} is the mass of the hydrogen atom, and R_{s} is the specific gas constant. We can compare H and H_{c} to estimate the vertical extent of clouds:

$$\boxed{\frac{H_{\mathrm{c}}}{H} = \frac{R_V}{R_{\mathrm{s}}} \frac{\mu c_p T_{\mathrm{c}}}{L}.} \tag{4.47}$$

For solar system planets the cloud to atmosphere scale height ratio can be estimated from the above equation to be about 0.05–0.2, meaning that clouds typically do not have a huge vertical extent. For example, as measured, Earth's scale height is about 8 km and clouds made of liquid water droplets have cloud scale heights of about 1.5 km. On Jupiter the numbers are about 43 km and 7 km, respectively. If a planetary atmosphere is typically 5–10 scale heights, we can see that cloud layers are relatively vertically thin. A more detailed table of planetary cloud condensates is given in Appendix B.

An upper limit for the cloud density can be obtained by assuming that all of the vapor condenses into liquid or solid form. The cloud density is the mass of the condensates divided by the total cloud volume.

In the ideal situation we would have a cloud formation model for predicting the cloud vertical and horizontal extent, density, and particle size distributions. Even for general circulation models of Earth, the representation of clouds is simplified from reality by parameterizing cloud-scale motions and cloud microphysical processes. More detailed, and computationally demanding, models of Earth clouds simulate 3D cloud-scale motions, resolve the size distribution of cloud droplets (and the aerosols on which they form) and treat the interactions between dynamics, microphysics, and radiative transfer [11]. These approaches are too detailed for exoplanets, where in situ measurements for basic cloud input parameters are impossible. Additionally, the exoplanet atmospheric dynamics and the populations of nuclei on which cloud droplets form are not known. Simpler models are therefore used for cloud formation in exoplanet atmospheres. The 1D models compare rates of particle growth (via processes such as convective uplifting of condensable vapor, nucleation, coagulation, and coalescence) and rates of particle destruction (primarily due to gravitational sedimentation). Again, the cloud formation model goal is to generate the fraction of gas condensed and cloud particle size as a function of altitude, assuming horizontally homogeneous clouds. See Figure 4.12 and [11, 12] and references therein for a detailed description of cloud formation models used for 1D exoplanet atmospheres, and [7] for cloud models and cloud particle interactions with radiation.

4.4.4 Cloud Particle Size Distributions

The size distribution of cloud particles is very significant for the magnitude of absorption and the magnitude and direction of scattering of radiation in a planetary atmosphere. Particle sizes are determined from in situ measurements on Earth and on planets that space probes have visited. Model fits to polarization data from Venus can also be used to infer the particle size.

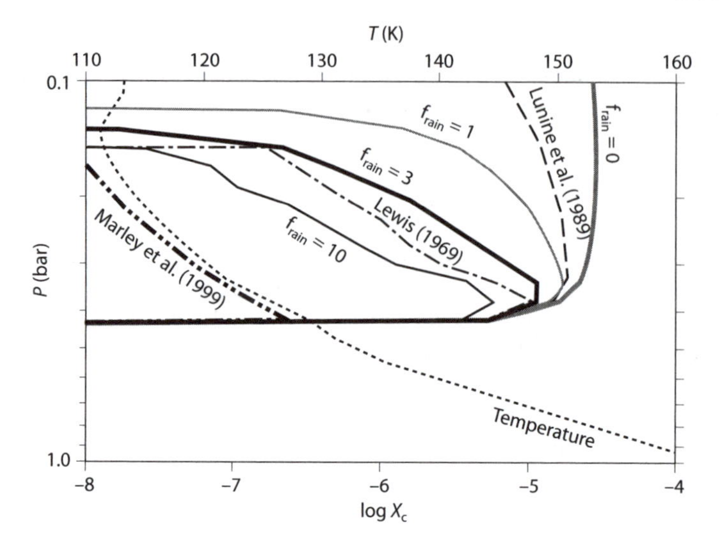

Figure 4.12 Cloud vertical extent and density for Jupiter's ammonia clouds. Shown is the condensate (i.e., cloud particle) mixing ratio (X_c) versus pressure. Different cloud formation models are shown. Adapted from [11] and see references therein.

The solar system planet cloud particle sizes span a wide range. Cloud particles in Venus have a size of 0.85–1.15 μm, and a haze layer above has particles with radius $r = 0.2$ μm [13]. The clouds on Jupiter range from an upper haze layer with $r = 0.5$ μm to lower cloud decks with $r = 0.75$ μm and $r = 0.45$–50 μm [14].

Cloud particles are not usually one fixed size. Instead, cloud particle sizes range over a distribution. To fit cloud particle size distributions people have used the gamma distribution, the log-normal distribution, and a power-law distribution. Earth's liquid water clouds have a size distribution that can be described by a log-normal distribution [15],

$$n(a) = \left(\frac{a}{a_0}\right)^6 \exp\left[-6\left(\frac{a}{a_0}\right)\right], \tag{4.48}$$

where a is the particle size and $a_0 \sim 4$ μm. Earth's stratospheric aerosols follow [15]

$$n(a) = \left(\frac{a}{a_0}\right) \exp\left[-2\left(\frac{a}{a_0}\right)^{1/2}\right]. \tag{4.49}$$

Entire books or parts of books are devoted to cloud formation and the optical effects of clouds and hazes. We will explore the opacity of cloud particles in Chapter 8.

4.5 ATMOSPHERIC ESCAPE

Atmospheric escape is significant for exoplanets in two major ways. First, atmospheric escape controls the bulk composition of a planet atmosphere or whether a planet even has an atmospere at all. Atmospheric loss usually refers to loss of light gases early in a planet's history, but may also refer to loss of heavier elements as relevant for terrestrial exoplanets orbiting very close to their host stars. Some hot super Earth exoplanets orbiting very close to the host stars, for example, may have lost their entire atmosphere over time. These hot super Earths are more massive analogs of atmosphereless Mercury. Other planets more similar to Earth may have had hydrogen gas in their atmospheres after formation but eventually lost the hydrogen via slow escape to space. Yet other hot exoplanets may be remnants of still more larger giant planets that had lost a significant fraction of their H and He envelope.

The second way that atmospheric escape is relevant for exoplanets is the continual loss rate of light gases. This loss rate provides a sink that can play an important role in photochemistry. On Earth, hydrogen is produced via photodissociation of hydrogen compounds (compounds such as H_2O and H_2S from volcanic emission). There is no net accumulation of hydrogen in Earth's atmosphere; Earth's hydrogen is primarily lost by atmospheric oxidation, with tiny amounts lost by Jeans escape, nonthermal charge-exchange reaction escape, and polar wind escape. While on Earth atmospheric escape is not the dominant loss rate for hydrogen, the situation could be different on an exoplanet. A key question is whether or not some super Earth exoplanets have hydrogen in their atmospheres. In exoplanets with oxygen-poor atmospheres and with higher escape velocities than Earth, hydrogen will accumulate in the atmosphere—highlighting the need for a careful consideration of photodissociation and atmospheric escape processes.

Atmospheric escape involves three major stages. The first stage is transport of gases from the lower to the upper atmosphere where escape can take place. The second stage is conversion from the atmospheric gas (usually in molecular form) to the escaping form (usually atomic or ionic). The third stage is the actual escape process itself. Any one of these stages can be the bottleneck, that is, the limiting process for atmospheric escape. Our goal is an overview of the escape processes. More details are covered in excellent review articles [16, 17, 18].

We can categorize atmospheric escape into three general processes, each described in a subsection below: thermal hydrostatic escape, thermal hydrodynamic escape, and nonthermal escape. Which escape process dominates on a given solar system planet depends on the planet's mass, the planet's upper atmosphere composition, the distance from the Sun, the consequent planetary exospheric temperature, and the presence of a magnetic field. The unknown properties of exoplanet atmospheres and the exoplanet host star make it difficult to understand or predict atmospheric escape on exoplanets.

The stellar type also affects atmospheric escape, because some star types (and some stars within a type) are more active than others. The EUV radiation from stars heats the upper atmosphere and increases atmospheric escape, yet EUV radiation for an individual star is difficult to measure without an ultraviolet-capable space telescope. Even a powerful EUV telescope will be limited to nearby stars, due to the absorption from the interstellar medium. The stellar wind also is not

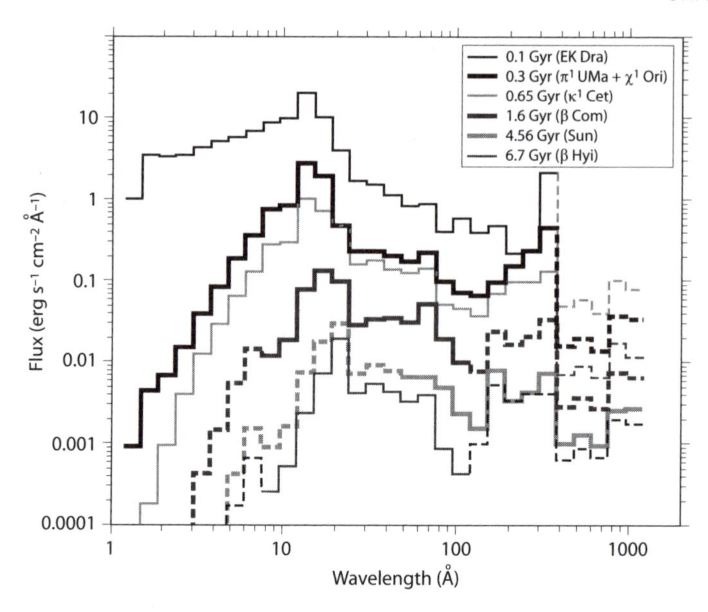

Figure 4.13 Spectral energy distribution of solar-type stars at different stages of main-sequence evolution. The solid lines are measured fluxes and the dotted lines are calculated fluxes interpolated from a power law relationship. Note the UV flux differences of up to 1000 times for the present-day Sun compared to a 100-million-year old sunlike star. This flux difference is in contrast to that at visible wavelengths, where the Sun was about 30% less luminous about 4 billion years ago. Adapted from [19].

measurable but can play a significant role in nonthermal escape mechanisms. To make matters even more complicated, the EUV and stellar wind histories that may have caused atmospheric loss early in a planet's evolution are typically not known for an individual star and have to be estimated (see Figure 4.13).

4.5.1 Thermal Escape (Hydrostatic Escape)

The most basic picture for atmospheric escape is that of thermal escape: an atom or molecule will escape a planet's atmosphere if its thermal velocity exceeds the escape velocity of the planet,

$$\boxed{\sqrt{\frac{2k_{\mathrm{B}}T_{\mathrm{exo}}}{m}} > \frac{1}{6}\sqrt{\frac{2GM_{\mathrm{p}}}{R_{\mathrm{p}}}}.} \tag{4.50}$$

Here, k_{B} is Boltzmann's constant, T_{exo} is the exospheric temperature, m is the molecular or atomic mass, G is the gravitational constant, M_{p} is the planet mass, and R_{p} is the planet radius.

Comparison of a molecule's thermal velocity with the planet escape velocity is not a completely realistic picture because the escaping particles are mostly those in the high-energy part of the velocity distribution. We have therefore multiplied by a

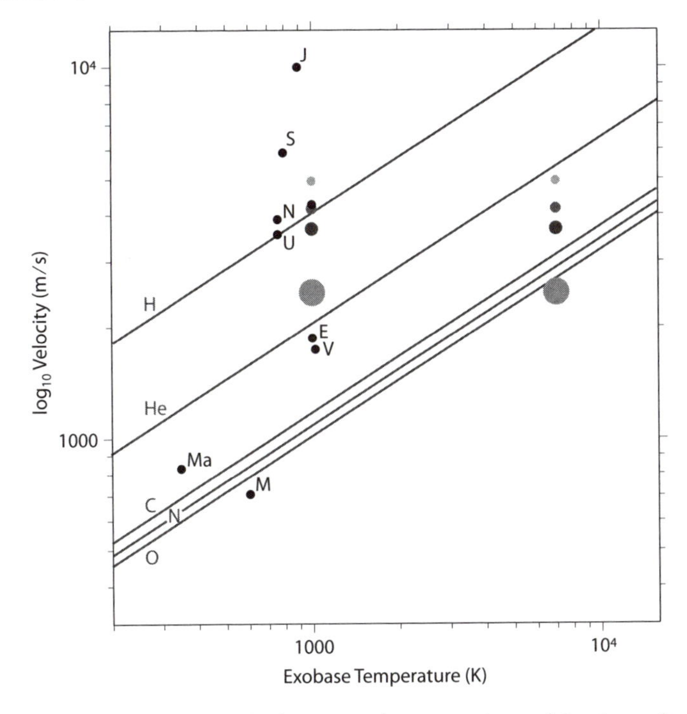

Figure 4.14 Illustration of atmospheric escape, by comparison of the thermal velocity of different atoms with 1/6 of the planet's escape velocity.

factor of 1/6 in equation [4.50] to estimate the escape of a gas species based on the velocities in the high-energy tail.

Loss of atmospheric particles by thermal escape occurs in the planetary exosphere, so the relevant temperature in equation [4.50] is the temperature in the exosphere. The exosphere is the uppermost region of the plantary atmosphere; the bottom of the exosphere is the exobase, defined to be the altitude at which the atmosphere becomes collisionless. In other words, the exobase is the altitude where a particle's mean free path (the average distance between particle collisions) is greater than the particle-specific atmospheric scale height (see equation [9.11] and the related discussion).

Despite its simplicity, the thermal escape criterion and the corresponding Figure 4.14 are useful in understanding a few basic elements of thermal atmospheric escape. First, equation [4.50] shows how atmospheric escape depends on mass. Second, equation [4.50] shows how some planets are not likely to lose their atmospheres, no matter how high the exospheric temperature.

A more accurate description of escape than a straight comparison of thermal velocity to escape velocity is the classical thermal Jeans escape. The standard view of Jeans escape is thermal "evaporation" from an exobase. In Jeans escape, the picture is that the thermal velocities of atoms and molecules of a given mass follow a Maxwellian distribution; Jeans escape describes the escape of molecules from the high-energy tail of the Maxwellian velocity distribution. Once they have escaped, slower atoms and molecules move to fill the high-velocity tail (this step may be diffusion limited). Recall the equation for the Maxwellian velocity distribution which gives the number of molecules with speeds between v and $v + dv$,

$$f(v)dv = \frac{4n}{\sqrt{\pi}} \left(\frac{m}{2k_{\rm B}T} \right)^{3/2} v^2 {\rm e}^{-mv^2/2kT} dv, \tag{4.51}$$

where m is the molecular mass, $k_{\rm B}$ is Boltzmann's constant, and T is the temperature. Here, n is the number density of particles. From the hydrostatic equilibrium equation derived in Chapter 9 and with a constant temperature the number density follows

$$n(r) = n(r_0){\rm e}^{-r/H(r_0)}, \tag{4.52}$$

where $H = kT/\mu_{\rm m} m_{\rm H} g(r)$ is the scale height derived in Section 9.3.3, equation [9.11]. The Jeans escape flux can be computed by integration over the direction and velocity appropriate for escaping particles

$$\Phi_{\rm Jeans} = \int_{v_{\rm esc}}^{\infty} \int_{0}^{2\pi} \int_{\pi/2}^{0} f(v) \cos\theta \sin\theta d\theta d\phi. \tag{4.53}$$

Here, integration is over the upper hemisphere for escaping particles, and it is assumed that the particles start at the exobase whereby they move on collision-free trajectores. We find the Jeans escape flux

$$\boxed{\Phi_{\rm Jeans} = \frac{n_{\rm c}}{2\sqrt{\pi}} B \sqrt{\frac{2k_{\rm B}T_{\rm c}}{m}} (1 + \lambda_{\rm c}){\rm e}^{-\lambda_{\rm c}},} \tag{4.54}$$

where the subscript c refers to exobase properties and the escape parameter $\lambda_{\rm c}$ is

$$\lambda_{\rm c} = \frac{GM_{\rm p}m}{kTr_{\rm c}} = \frac{r_{\rm c}}{H}. \tag{4.55}$$

The escape parameter can also be described by $\lambda_{\rm c} = E_{\rm esc}/kT_{\rm c}$ where the escape energy $E_{\rm esc} = \frac{1}{2}mv_{\rm esc}^2$. Here $M_{\rm p}$ is the planetary mass, r is the radial distance from the planet center, and G is the universal gravitational constant. $\phi_{\rm Jeans}$ is the escape flux in units of cm^{-2}s^{-1}. The factor B accounts for the slow repopulation time of the energetic tail of the Maxwellian distribution and has a value of the order 0.5–0.8. To determine if a given gas species is escaping by Jeans escape one has to compare the Jeans escape flux to other escape fluxes. If Jeans escape is the dominant escape mechanism, one can determine if an atmospheric gas has been depleted by Jeans escape by taking the original amount of the given gas species in the atmosphere, computing the depletion time, and also considering production rates (if any) of the gas species in question.

4.5.2 Hydrodynamic Escape

Hydrodynamic escape occurs when the atmospheric escape is so fast that the atmosphere behaves like a dense fluid expanding radially outward. The hydrodynamic escape state is reached when the atmosphere is heated by the star, to the point where the particles are so energetic that gravity cannot stop the outward flow. For hydrodynamic escape to occur, a very large amount of EUV radiation is needed to power the outward atmosphere flow against gravity; typically very high exospheric temperatures and corresponding very high levels of stellar EUV fluxes are required. As for thermal hydrostatic escape, the relevant part of the stellar flux is the EUV flux, because the EUV flux gets absorbed high in the atmosphere and drives heating of the uppermost atmosphere layers.

During hydrodynamic escape, heavier elements such as C, N, and O can be carried away with the hydrodynamic flow of the lighter species H. Depending on the atmospheric composition, heavy elements themselves might be the main constituents of the hydrodynamic flow, not depending on the H flow for hydrodynamic escape. Hydrodynamic escape is typically associated with an episode of atmospheric escape early in a planetary history. Pluto is probably the only solar system body undergoing signficant atmospheric hydrodynamic escape at present (of CH_4 and N_2) [20].

To estimate whether an atmosphere is in the hydrostatic or hydrodynamic escape regime we can use the Jeans escape parameter λ_c. Typically the value $\lambda_c \approx 1$ is the dividing point, with $\lambda_c \gg 1$ safely in the Jeans escape regime. In this sense hydrodynamic escape can be considered as an asymptotic case of thermal escape. Small values of λ_c are necessary but not sufficient to drive hydrodynamic flow. Model calculations are needed to understand the escape regime. In more detail, the transition between the hydrostatic and hydrodynamic regime is controlled by adiabatic cooling in the energy budget. The more energy put into the planet atmosphere, the greater the escape rate; the faster the flow, the more energy is consumed. Eventually, a negative temperature gradient at the top of the atmosphere, near the exobase develops because the stellar radiation is not able to overcome the cooling rate. This decreasing (rather than constant) temperature with increasing altitude would be a macroscopic signature of hydrodynamic escape, if it were observable.

An estimate for hydrodynamic escape is the so-called energy-limited escape. For exoplanets, the energy-limited escape is useful to illustrate the main uncertainties in EUV-driven escape [21]. The gravitational potential energy of the planet in J is

$$E_{\mathrm{p}} = -\frac{GM_p m_{\mathrm{atm}}}{\beta R_{\mathrm{p}}}, \tag{4.56}$$

where G is the gravitational constant, M_{p} is the planet mass, R_{p} is the planet radius, and βR_{p} is the radial extent of the planet exosphere. Here m_{atm} is the mass of the planet atmosphere, which we assume to be small enough that we can ignore the radial atmosphere structure in our estimate. The EUV power [J s^{-1}] incident on the planet is

$$P_{UV} = \pi R_{\mathrm{p}}^2 F_{\mathrm{EUV}}. \tag{4.57}$$

We assume here that the radius where the EUV energy is absorbed is close enough to the planetary radius. The ratio of equations (4.56) and (4.57), and an active-phase average EUV flux $\langle F_{\mathrm{EUV}} \rangle$ lead to the lifetime of the initial planet atmosphere

$$\tau \sim \left[\frac{G}{\pi}\right] \left(\frac{M_{\mathrm{p}}}{R_{\mathrm{p}}^3}\right) \frac{m_{\mathrm{atm}}}{\beta \eta \langle F_{\mathrm{EUV}} \rangle}. \tag{4.58}$$

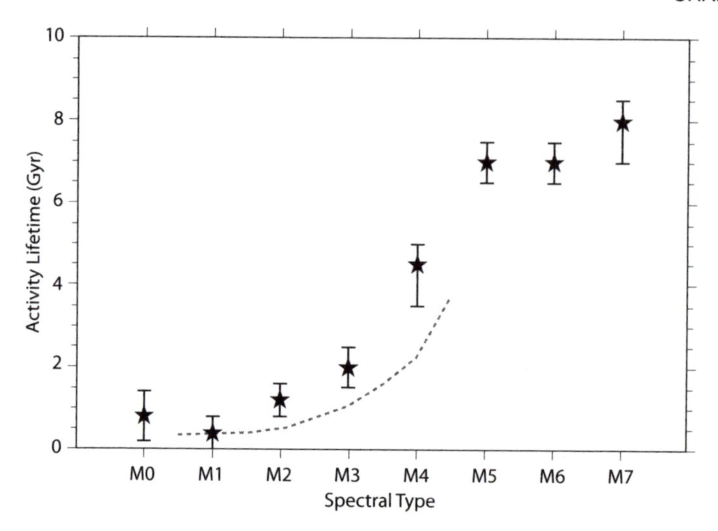

Figure 4.15 Stellar activity lifetimes for M stars. While solar-type stars (not shown) are active for at most a few hundred million years, M stars can be active for billions of years. The strong EUV radiation during the active phase can drive thermal atmospheric escape on exoplanets. Adapted from [24].

Energy-limited escape assumes that some fraction η of the incident UV flux heats the atmosphere and drives escape. Note that $\beta > 1$ and $\eta < 1$. This equation is valid provided that the upper atmosphere is heated strongly enough for the relevant elements (e.g., H or C) to escape.

We can see the main uncertainties in atmospheric thermal escape from equation [4.58], even though the equation is approximate at best. First, if a planet is transiting we know the planetary density. If the planet is not transiting and we assume the planet is solid, the density varies by less than a factor of 5 for reasonable interior compositions (Mercury-like to a "dirty water" planet) and for a range of planetary masses. More uncertain is the initial outgassed planetary atmosphere mass, which could vary widely, from less than 1% of the planet's total mass up through 5% and even as high as 20% for an initially water-rich planet [22]. An equally large unknown is the incident EUV flux. All stars have very high EUV flux during their so-called saturation phase, where levels can reach thousands of times present-day solar EUV levels (Figure 4.13). The evolution of the solar EUV radiation with time is thought to follow $(t_0/t)^{5/6}$, where t_0 is the present time at 4.6 billion years [23]. The high EUV is most critical for M star super Earths: M star saturation phases can last 1–3 billion years for early M stars and 6–8 billion years for late M stars (Figure 4.15). If the initial planetary atmosphere has not survived during the star's saturation phase, a new atmosphere could develop in the star's quiet phase only if the outgassing rate is higher than the escape rate, and there is enough interior material to outgas to form an atmosphere.

Actual calculations for hydrodynamic escape are based on the hydrodynamic equations of fluid mechanics. The calculations require the solar or stellar EUV

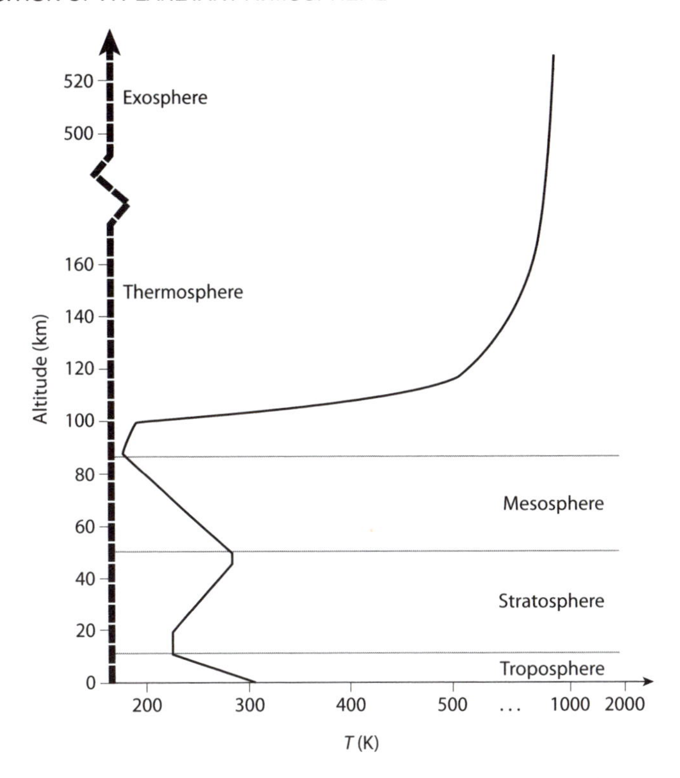

Figure 4.16 Earth's temperature profile showing the thermosphere and the exosphere. The exosphere extends to an altitude 500 km above Earth's surface. The ionosphere is a layer within the thermosphere. The temperature profile is from Earth's standard reference atmosphere and is based on measurements.

radiation and masses and constituents of planetary upper atmospheres. These inputs can be challenging for early planetary atmospheres and for exoplanets and their host stars, simply due to lack of information for input values. The calculations yield escape fluxes and density and temperature structure of the thermosphere and exosphere (Figure 4.16).

4.5.3 Nonthermal Escape

"Nonthermal escape mechanisms" generally refer to collisional processes between charged species that produce atoms energetic enough to escape from a planetary atmosphere to space. These collisional processes enable the escape of heavier atoms such as N, C, and O, whose thermal velocities are too small for thermal escape. Nonthermal escape can also refer to sputtering and stellar wind erosion. Atoms that do not fully escape the planet may end up in a hot planetary "corona." In general, a wide variety of processes are included under the nonthermal escape umbrella. Which nonthermal process dominates or even plays an important role depends on many different planet-specific factors, including the exospheric temperature, the

Table 4.1 Atmospheric escape processes.

Process	Examples	Planet (gas) example
Charge exchange	$H + H^{+*} \rightarrow H^{+} + H^{*}$	Earth (H, D)
	$O + H^{+*} \rightarrow O^{+} + H^{*}$	Venus (He)
Dissociative recombination	$O_2^{+} + e \rightarrow O^{*} + O^{*}$	Mars (O), E, G, C (O)
	$OH^{+} + e \rightarrow O + H^{*}$	Venus (H), Mars (N), Titan (H_2)
Impact dissociation	$N_2 + e^{*} \rightarrow N^{*} + N*$	Mars (N), Titan (N)
Photodissociation	$O_2 + h\nu \rightarrow O^{*} + O^{*}$	
Ion-Neutral reaction	$O^{+}H_2 \rightarrow OH^{+} + H^{*}$	
Sputtering or knockon	$Na + S^{+} \rightarrow Na^{*} + S^{+*}$	Io (Na, K)
	$O^{*} + H \rightarrow O^{*} + H^{*}$	Venus (H)
Solar-wind pickup	$O + h\nu \rightarrow O^{+} + e$ then O^{+} picked up	Mercury (He, Ar)
Ion escape	H^{+*} escapes	Earth (H, D, He)
Jeans escape		Earth (H, D), Mars (H, H_2), Titan (H, H_2), Pluto (CH_4)

The * represents excess kinetic energy. The Jeans escape is a thermal process but is included in this table for completion. Adapted from [16].

presence of a planetary magnetic field, the planet-star separation, the host star type, and especially the planet's escape velocity.

In this section we provide an overview by listing examples of nonthermal escape (see Table 4.1). For more details on nonthermal escape processes and a summary of numerical models see the excellent review articles in [16,17].

Charge exchange involves a collision between an ion and an atom, where the excess energy of the ion is transferred to the atom. The solar EUV radiation generates ions via photodissociation of atoms. The ions get their high energies from the magnetosphere. Charge exchange from the H^{+} ion to the H atom is thought to be the dominant hydrogen loss process, even faster than Jeans escape of hydrogen. Charge-exchange reactions also involve energetic protons and other thermal atoms such as oxygen.

Conversion of photochemical energy into kinetic energy. Electrons created from photoionization or atoms created from molecular photodissociation can carry excess kinetic energy. Additionally, atoms and ions generated from photodissociation will recombine, and impacts can yield excess kinetic energy of a few eV. Dissociative recombination occurs when a molecular ion combines with a free electron, generating a population of energetic atoms. Photochemical escape is only important for present-day Mars, not for Venus or Earth.

Ion escape refers to ions that escape along magnetic field lines that are open to a planetary magnetotail. On Earth ion escape is an important process in the polar regions; hence this nonthermal escape process is also referred to as the polar wind.

Sputtering and knockon. Impacts of atoms or ions onto a surface generate a "back splash" of atoms; this process is known as sputtering. The surface could be either a solid surface, or a layer at the top of the atmosphere. In such a layer, the impacts may generate forward acceleration; this process is called knockon. Knockon leads to atmospheric escape by a subsequent cascade of collisions that eventually generate energetic particles heading back out in the direction to escape the planet's atmosphere.

Pickup of ions happens when the solar wind interacts with a planet's ionosphere. The Sun's magnetic field extends to the planets by the charged particles in the solar wind, and is referred to as the "interplanetary magnetic field." The ions in the planet atmosphere are "picked up" by the solar interplanetary magnetic field and can be accelerated away from the planet. Planetary magnetic fields, such as Earth's, protect the atmosphere from escape by deflecting the solar wind. In contrast, in planets that lack a magnetic field, the solar wind can penetrate deep into the exosphere. There, the solar wind plasma can ionize neutral species and sweep them away.

4.6 ATMOSPHERIC EVOLUTION

Our main goal in this book is to understand the origin and interpretation of atmospheric spectra. Nevertheless, it is important to recognize the significant atmospheric evolution that has brought the solar system terrestrial planet atmospheres—Earth, Mars, and Venus—to their present state.

Earth's atmospheric evolution is arguably the most dramatic of all of the solar system planets. Earth's atmosphere, like the other terrestrial planets, was formed from volcanic outgassing during shortly after formation. Earth's early atmosphere was dominated by CO_2 and N_2, but outgassing also produced H_2O vapor and some NH_3, CH_4, and H_2. While the amount of hydrogen present in the early Earth's atmosphere is not known, some researchers have speculated that hydrogen may have reached up to 30% of the atmosphere by volume. Earth's atmospheric hydrogen was lost by atmospheric escape processes and so any such excess was probably gone by hundreds of millions of years or much sooner after formation, although the exact duration of any hydrogen-rich phase is unknown. Earth's atmosphere continued to have roughly 100–1000 ppmv of H_2 until much later as a consequence of ongoing outgassing of H_2 and other reduced gases, a process that continues even today.

Much of Earth's initial CO_2 was dissolved in the ocean and precipated out as carbonates, eventually transforming Earth's atmosphere from one dominated by CO_2 to an atmosphere containing less than 1% of CO_2. The rise of O_2 on Earth, beginning $\sim$2.4 billion years ago, dramatically changed Earth's atmosphere forever. Oxygen is produced in large quantities only by life. Cyanobacteria, the first life to develop photosynthesis began producing oxygen about 300 million years before Earth's atmosphere became oxygenated. The O_2 atmospheric concentration likely remained much lower than its present-day levels of 21% until at least 0.6 billion years ago. The oxygen did not initially accumulate in the atmosphere, because the oxygen first produced was used up in reactions with Earth's surface rocks.

The presence of N_2 in Earth's atmosphere is somewhat of a mystery. N_2 is, however, largely inert and is likely left over from outgassed ammonia in Earth's past. N_2 should also have been produced in Earth's atmosphere by the impact plumes of C- and N-rich impactors. Today, Earth's atmosphere consists of 20.9% oxygen, 78.1% nitrogen gas, and 0.93% Ar, with the rest of the atmosphere composed of other gases including H_2O and CO_2.

Venus has a remarkably different atmosphere from Earth's, despite the planets having approximately the same size and mass. In its distant past, however, Venus is thought to have had not only an early atmosphere similar to Earth's early atmosphere but also a surface water ocean tens to hundreds of meters deep. What happened? Venus is thought to have suffered a "runaway greenhouse," in which water vapor from the ocean evaporated into the atmosphere, was subsequently split up by solar EUV photons, and the hydrogen escaped to space. The total hydrogen content could have been lost through hydrodynamic escape in less than 2×10^9 years [25]. If the hydrogen from H_2O escaped, what happened to the oxygen? At present there is no significant amount of molecular oxygen in the Venusian atmosphere. Oxygen might have been dragged off with the hydrogen hydrodynamic flow, or could have been removed from the atmosphere by reaction with reduced minerals in Venus's crust. Venus's lack of a magnetic field means that the solar wind (stronger by 30–1000 times during the Sun's first billion years) could have potentially enhanced the hydrodynamic escape flux. At present, without an ocean to sequester CO_2, Venus has almost 100 times more CO_2 in its atmosphere than does Earth. The differences between present-day Venus and Earth atmospheres originated with atmospheric escape due to Venus's closer distance to the Sun and Venus's lack of a strong magnetic field.

Mars's atmosphere also has changed dramatically during its first 0.5 to 1 billion years. Owing to its small size and mass and low surface gravity, present-day Mars hardly has an atmosphere at all. At 1.5 AU from the Sun, Mars's present-day surface temperature is too cold to support liquid water. Yet there is considerable strong evidence that early Mars had liquid surface water—and therefore must have had an atmosphere to keep the planet warm. The atmosphere must have been fairly massive (1–3 bars vs. the 6 mbar today) and contain a significant amount of greenhouse gases (e.g., CO_2), although a CO_2-H_2O greenhouse alone cannot have kept early Mars's atmosphere above freezing [26].

There is no question that massive atmosphere loss on early Mars was possible—due to Mars's low surface gravity and lack of a substantial magnetic field after 700 million years [27], and the stronger solar EUV flux and solar wind during the Sun's first billion years. Atmosphere erosion following the impact of a massive planetesimal may also have played a role. The timescale and responsible processes for Mars's atmospheric loss require models of the early solar EUV radiation and wind, models of the Martian early upper atmosphere, and an adopted history of Mars's magnetic field.

Exoplanets have masses and planet-star separations with almost all imaginable values (Figure 1.5). The resulting range of possibilities for atmospheric escape is therefore enormous. It is challenging to get a handle on atmospheric escape for a given exoplanet due to many unknown facts: present-day atmospheric composition, history of the host star, the original atmosphere mass and composition, and whether or not the planet has a magnetic field. As we have seen there are a wide variety of atmospheric escape mechanisms. The imprint on solar system planet atmospheres is in the observed patterns of isotopic fractionation of noble gases, and these imprints are used to try to determine escape histories—measurements not obtainable for exoplanets.

Observations that are accessible for exoplanets include those of the bulk composition of exoplanet lower atmospheres and the presence of escaping gases that fill a large, extended exosphere. The exoplanet HD 209458b must have a huge extended exosphere, based on observations of neutral atomic hydrogen, seen in Lyman α absorption during transit [28]. Models estimate the escape rate of hydrogen up to about a few $\times 10^{10}$ g s^{-1}, implying that less than 1% of the planet's mass would be lost over the planet's lifetime. Perhaps it seems puzzling that hot Jupiters—some at ten times closer to their star than Mercury is to the Sun—can maintain their atmospheres at all. The big picture answer is that the hot Jupiters are massive (hundreds of times Earth's mass) so that they are able to hold on to their massive atmospheres, despite heating by the EUV radiation from their host stars. People have speculated that many hot Jupiters may have lost significant fractions of their initial mass and furthermore that some hot Neptune-mass exoplanets or even hot super Earths may be remnant cores of more massive planets whose atmospheres or envelopes were completely lost.

4.7 SUMMARY

We have described where the chemical composition of an exoplanet atmosphere comes from. We started with a presentation of Jupiter's and Earth's atmospheric composition. We then used chemical equilibrium to describe atomic and molecular mixing ratios, using the elemental abundances, temperature, and pressure in a given planetary atmosphere. For giant planets with deep atmospheres that reach high temperatures, chemical equilibrium provides an adequate description of most of the constituents of the atmosphere. Some gas species, however, can still depart significantly from chemical equilibrium. These gas species (such as CO and N_2) get vertically mixed upward from the deep atmosphere where they are favored and take too long to be converted to the chemical equilibrium form in the upper atmosphere. In this case the timescale for vertical mixing and the timescale for chemical equilibrium of a given gas species must be considered. Beyond chemical equilibrium, photochemistry is necessary to estimate the atmospheric composition for terrestrial planets whose surface gravity makes atmospheric escape easier than for giant planets. Atmospheric evolution is signficant for terrestrial-like planets, but mostly happens in the first hundreds of millions to billions of years. In this book we focus on the planet atmosphere after early evolution is over.

REFERENCES

References for this chapter

1. Beichman, C. A., Woolf, N. W., and Lindensmith, C. A., eds. 1999. *Terrestrial Planet Finder* JPL Publication 99-003.

2. Anders, E., and Grevesse, N. 1989. "Abundances of the Elements: Meteoritic and Solar." Geo. Cos. Acta 53, 197–214.

3. Cox, A.N.. *Allen's Astrophysical Quantities* (4th ed; New York: Springer).

4. Karkoschka, E. 1994. "Spectrophotometry of the Jovian Planets and Titan at 300- to 1000-nm Wavelength: The Methane Spectrum." Icarus 111, 174–192.

5. Marley, M., Fortney, J., Seager, S., and Barman, T. 1997. "Atmospheres of Extrasolar Giant Planets," in *Protostars and Planets V*, eds. V. Reipurth, D. Jewitt, and K. Keil (Tucson: University of Arizona Press).

6. Saumon, D., Marley, M. S., Cushing, M. C., Leggett, S. K., Roellig, T. L., Lodders, K., and Freedman, R. S. 2006. "Ammonia as a Tracer of Chemical Equilibrium in the T7.5 Dwarf Gliese 570D." Astrophys. J. 541, 374–389.

7. Liou, K. N. 2002. *An Introduction to Atmospheric Radiation* (London: Academic Press).

8. Pont, F., Knutson, H., Gilliland, R. L., Moutou, C., and Charbonneau, D. 2008. "Detection of Atmospheric Haze on an Extrasolar Planet: the 0.55-1.05 μm Transmission Spectrum of HD 189733b with the *Hubble Space Telescope*." Mon. Not. R. Astron. Soc. 385, 109–118.

9. Lodders, K. 2004. "Brown Dwarfs–Faint at Heart, Rich in Chemistry." Science 303, 323–324.

10. Sanchez-Lavega, A., Perez-Hoyos, S., and Hueso R. 2004. "Clouds in Planetary Atmospheres: A Useful Application of the Clausius-Clapeyron Equation." Am. J. Phys. 72, 767–774.

11. Ackerman, A. S., and Marley, M. S. 2001. "Precipitating Condensation Clouds in Substellar Atmospheres." Astrophys. J. 556, 872–884.

12. Cooper, C. S., Sudarsky, D., Milsom, J. A., Lunine, J. I., and Burrows, A. 2003. "Modeling the Formation of Clouds in Brown Dwarf Atmospheres." Astrophys. J. 586, 1320–1337.

13. Knibbe, W. J. J., de Haan, J. F., Hovenier, J. W., and Travis, L. D. 1997. "A Biwavelength Analysis of Pioneer Venus Polarization Observations." J. Geophys. Res. 102, 10945–10958.

14. Taylor, F., and Irwin, P. 1999. "The Clouds of Jupiter." Astron. Geophys. 40, 21.

15. Deirmendjian, D. 1964. "Scattering and Polarization Properties of Water Clouds and Hazes in the Visible and Infrared." Appl. Optics. 3, 187–196.

16. Hunten, D. M. 1982. "Thermal and Nonthermal Escape Mechanisms for Terrestrial Bodies." Planet. Space Sci. 30, 773–783.

17. Shizgal, B. D., and Arkos, G. G. 1996. "Nonthermal Escape of the Atmospheres of Venus, Earth, and Mars." Rev. Geophys. 34, 483–505.

18. Chassefiere, E., and Leblanc, F. 2004. "Mars Atmospheric Escape and Evolution; Interaction with the Solar Wind." Planet. Space Sci. 52, 1039–1058.

19. Ribas, I., Guinan, E. F., Güdel, M., and Audard, M. 2005. "Evolution of the Solar Activity over Time and Effects on Planetary Atmospheres. I. High-Energy Irradiances (1–1700 Å)." Astrophys. J. 622, 680–694.

20. Stern, S. A. 1992. "The Pluto-Charon System." Ann. Rev. Astron. Astrophys. 30, 185–233.

21. Lecavelier Des Etangs, A. 2007. "A Diagram to Determine the Evaporation Status of Extrasolar Planets." Astron. Astrophys. 461, 1185–1193.

22. Elkins-Tanton, L., and Seager, S. 2008. "Ranges of Atmospheric Mass and Composition of Super Earth Exoplanets." Astrophys. J. 685, 1237–1246.

23. Zahnle, K. J., and Walker, J. C. G. 1982. "The Evolution of Solar Ultraviolet Luminosity." Rev. Geophys. 20, 280–292.

24. West, A. A., Hawley, S. L., Bochanski, J. J., Covey, K. R., Reid, I. N., Dhital, S., Hilton, E. J., and Masuda, M. 2008. "Constraining the Age-Activity Relation for Cool Stars: the Sloan Digital Sky Survey Data Release 5 Low-Mass Star Spectroscopic Sample." Astron. J. 135, 785–795.

25. Kasting, J. F., and Pollack, J. B. 1983, "Loss of Water from Venus. I - Hydrodynamic Escape of Hydrogen." Icarus 53, 479–508.

26. Kasting, J. F. 1991, "CO_2 Condensation and the Climate on Early Mars", Icarus, 94, 1–13.

27. Acuna, M. H., et al. 1998. "Magnetic Field and Plasma Observations at Mars: Initial Results of the Mars Global Surveyor Mission." Science 279, 1676–1680.

28. Vidal-Madjar, A., Lecavelier des Etangs, A., Desert, J.-M., Ballester, G. E., Ferlet, R., Hebrard, G., and Mayor, M. 2003. "An Extended Upper Atmosphere Around the Extrasolar Planet HD 209458b." Nature 422, 143–146.

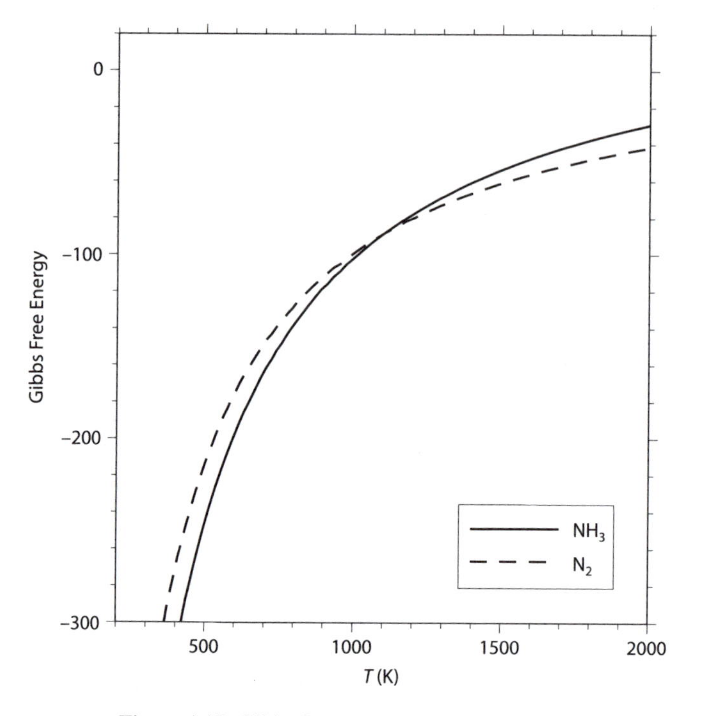

Figure 4.17 Gibbs free energies of NH_3 and N_2.

EXERCISES

1. Derive the Gibbs free energy equation (equation [4.11]) from the first law of thermodynamics (equation [4.3]) by filling in any missing steps.

2. Figure 4.17 shows the Gibbs free energy of the nitrogen compounds NH_3 and N_2. Can you use Figure 4.17 to determine which nitrogen molecule will dominate at a given temperature and pressure? Why or why not?

3. Compute the escape velocity of the Earth, Venus, and Mars, in m/s and in eV. Based on their exobase temperatures compute the escape energies of H and O, as well as the Jeans escape parameter λ_c. Should H and O thermally escape on Earth, Venus, or Mars?

4. Make a plot of relative stellar heating (in units of the solar luminosity at Earth, $L_\oplus = L_\odot/4\pi a_\oplus^2$) ($y$-axis) vs. escape velocity (x-axis) for all solar system planets, Pluto, Europa, Titan, Triton, and the Moon. On the same graph, plot all transiting exoplanets (take data from the Extrasolar Planet Encyclopaedia at http://exoplanet.eu/catalog.php, but consider only exoplanets with measured masses and radii). Because star luminosity data are not available, assume that $L/L_\odot = (M/M_\odot)^{3/5}$. Use a log-log plot. Comment on your findings.

Chapter Five

Radiative Transfer I: Fundamentals

5.1 INTRODUCTION

Radiative transfer describes how radiation changes as it travels through a medium. A wide variety of absorption and emission processes from atoms, molecules, and solid particles contribute to the radiation that emerges at the top of the planetary atmosphere. Although we can observe only the radiation emerging from the planet's atmosphere, much information is contained in the emergent spectrum, including details of the atmospheric temperature, pressure, and composition.

The radiative transfer equation describes the change in a beam of radiation as it travels some distance s through a volume of gas. The changes are due to losses from the beam $\kappa(\mathbf{x}, \nu)I(\mathbf{x}, \hat{\mathbf{n}}, \nu)$, where κ is the extinction coefficient and I is the intensity, and additions to the beam $\varepsilon(\mathbf{x}, \hat{\mathbf{n}}, \nu)$, where ε is the emission coefficient. Adding up the losses and gains, for an atmosphere that is not changing with time (i.e., static), the radiative transfer equation is

$$\boxed{\frac{dI(\mathbf{x}, \hat{\mathbf{n}}, \nu)}{ds} = -\kappa(\mathbf{x}, \nu)I(\mathbf{x}, \hat{\mathbf{n}}, \nu) + \varepsilon(\mathbf{x}, \hat{\mathbf{n}}, \nu).} \tag{5.1}$$

In this chapter we will omit the t-dependence of the intensity, flux, and other terms. Nevertheless, we must still keep in mind that radiation involves energy flow: energy per unit time.

The goal of this chapter is to lay the foundation for the equation of radiative transfer. We first introduce several fundamental quantities and concepts needed to characterize the radiation field. Next we decribe the full equation of radiative transfer and derive the plane-parallel approximation. We finish with an outline of the formal solution to the radiative transfer equation. We leave details of the solutions of the equation of radiative transfer until Chapter 6.

5.2 OPACITY

5.2.1 Definition of Opacity

The most important physical quantity in the atmosphere that affects the transfer of radiation is the opacity. The opacity describes how opaque a substance is: how hard it is for radiation to pass through that substance. Opacity depends on the number density of particles in the atmosphere, and the particles' absorbing, scattering, and emitting properties, which may in turn depend on temperature, pressure, and frequency (i.e., wavelength). In order to describe radiative transfer we therefore must define coefficients for the absorption and emission of radiation. The word opacity refers to absorption and scattering (i.e., extinction) coefficients of atmospheric particles. We emphasize that the underlying physical processes of photon absorption and emission, will, for now, be hidden in these macroscopic coefficients. The microscopic details of the absorption and emission coefficients will be covered in Chapter 8.

Before we describe the opacity in more detail, we should keep in mind that many different symbols for the opacity coefficients are used in other textbooks and various papers in the literature.

5.2.2 The Extinction Coefficient: Absorption and Scattering

The monochromatic extinction coefficient $\kappa(\mathbf{x}, \nu)$ describes the amount of energy removed from a beam of radiation from a volume dV and a solid angle $d\Omega$, per unit time per unit frequency,

$$\boxed{dE = \kappa(\mathbf{x}, \nu) I(\mathbf{x}, \hat{\mathbf{n}}, \nu) dV d\Omega d\nu dt.} \tag{5.2}$$

See Chapter 2 and Figure 2.1 for a description of the intensity I.

We use κ (with units of m^{-1}) to include all processes that remove energy from a beam of radiation. True absorption describes processes that destroy the photon. This can happen, for example, when a photon is absorbed by a particle, and then is converted to kinetic energy of the gas by a subsequent collision. We will denote the true absorption coefficient by $\alpha(\mathbf{x}, \nu)$. Scattering describes photons that are removed from the beam by a change of direction (and possibly a change of energy). We denote the scattering coefficient by $\sigma_{\mathrm{s}}(\mathbf{x}, \nu)$, where we use the subscript s to distinguish the scattering coefficient σ_{s} in m^{-1} from the cross section σ in m^2. The extinction coefficient is therefore comprised of

$$\kappa(\mathbf{x}, \nu) = \alpha(\mathbf{x}, \nu) + \sigma_{\mathrm{s}}(\mathbf{x}, \nu). \tag{5.3}$$

Extinction is isotropic in a static medium because the absorbing or scattering particle does not care which direction the photon is coming from. Hence κ has no $\hat{\mathbf{n}}$-dependence.

5.2.3 The Emission Coefficient: Thermal Emission and Scattering

The monochromatic emission coefficient $\varepsilon(\mathbf{x}, \hat{\mathbf{n}}, \nu)$ describes the amount of energy emitted into a volume dV within a solid angle $d\Omega$ per unit time, per unit frequency,

$$\boxed{dE = \varepsilon(\mathbf{x}, \hat{\mathbf{n}}, \nu) dV d\Omega dt d\nu,} \tag{5.4}$$

with dimensions J m^{-3} sr^{-1} s^{-1} Hz^{-1}. Planetary atmosphere emission includes both thermal emission and scattering, because both processes can add to the beam of radiation.

Thermal emission can be described by the Kirchhoff-Planck relation also known as Kirchhoff's Law of thermal radiation, or Kirchhoff's Law for short,

$$\varepsilon_{\mathrm{therm}}(\mathbf{x}, \nu) = \alpha(\mathbf{x}, \nu) B(\mathbf{x}, \nu), \tag{5.5}$$

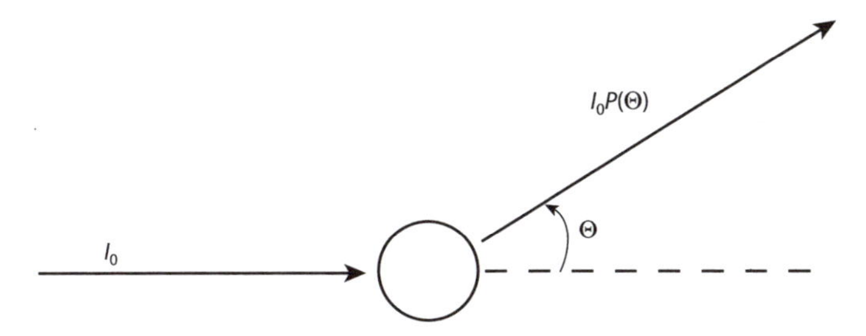

Figure 5.1 Definition of scattering angle used in $P(\Theta)$.

where we recall that $B(\mathbf{x}, \nu)$ is black body radiation, and $\mathbf{x}$ depends on temperature. This equation for thermal radiation is valid in local thermodynamic equilibrium, and is further described in Section 5.4. The thermal emission is isotropic in a static medium because atoms and molecules have no preferred direction of radiation.

The emission coefficient for pure scattering is

$$\varepsilon_{\mathrm{scat}}(\mathbf{x}, \hat{\mathbf{n}}, \nu) = \sigma_{\mathrm{s}}(\mathbf{x}, \nu) \frac{1}{4\pi} \int_{\Omega} P(\Theta) I(\mathbf{x}', \hat{\mathbf{n}}', \nu) d\Omega'. \tag{5.6}$$

Here, Θ is the scattering angle, $\Theta = \hat{\mathbf{\Omega}} \cdot \hat{\mathbf{\Omega}}'$, where in general the terms with primes refer to the direction of incidence and the terms without primes refer to the direction after scattering (Figure 5.1 and see ahead to a related discussion in Section 8.5). Here $\sigma_{\mathrm{s}}(\mathbf{x}, \nu)$ is the scattering coefficient, with dimensions m^{-1}. In this book we will consider only coherent scattering, the case where the photon retains the same energy (i.e., frequency).

Scattering can be anisotropic and so we define $P(\Theta)$ as the dimensionless single scattering phase function. The phase function denotes the redirection in the incident intensity to the outgoing intensity and describes the 3D directional scattering probability. The phase function is normalized to 1. (P should not be confused with the illumination phase function $\Phi(\alpha)$ used in Section 3.4.3 to describe observations.) We will describe the origin of the single scattering phase function in Chapter 8. The phase function is normalized:

$$\frac{1}{4\pi} \int_{\Omega} P(\Theta) d\Omega = 1. \tag{5.7}$$

If the scattering is isotropic, $P(\Theta) = 1$, and the emission coefficient for scattering (equation [5.6]) reduces to

$$\varepsilon_{\mathrm{scat}}(\mathbf{x}, \nu) = \sigma_{\mathrm{s}}(\mathbf{x}, \nu) J(\mathbf{x}, \nu). \tag{5.8}$$

Recall that $J(\mathbf{x}, \nu)$ is the mean intensity (equation [2.2]).

Before ending our description of scattering we emphasize two points. First, scattering is both a source and a sink for the beam of radiation in the planet atmosphere. As a source, scattering is counted as emission. As a sink, scattering is counted as extinction. Second, scattering is angle dependent, in contrast to the isotropic thermal emission. Hence the emission coefficient dependency on $\hat{\mathbf{n}}$.

5.2.4 Number Densities, Cross Sections, and the Absorption Coefficient

The absorption and scattering coefficients are the sum of the number density (n, in units of m^{-3}) of particles times their absorption or scattering cross sections (σ, in units of m^2). The number densities of different gas or solid species originate from elemental abundances in the planetary atmosphere and from atmospheric chemistry and escape (Chapter 4).

The number densities and absorption cross sections come into play via the absorption coefficient

$$\boxed{\alpha(T, P, \nu) = \sum_j \alpha_j(T, P, \nu) = \sum_j \sum_i n_{ji}(T, P)\sigma_{ji}(T, P, \nu).} \tag{5.9}$$

Here j refers to different atomic and molecular species and i refers to different atomic or molecular states (i.e., energy level populations). We note that, while we have previously used the location $\mathbf{x}$ to describe the absorption coefficient in an atmosphere (by $\alpha(\mathbf{x}, \nu)$), for a general description we prefer to use the T, P dependencies. Recall that the temperature T and pressure P vary with location $\mathbf{x}$ in an atmosphere; therefore for the extinction coefficient we can consider $\mathbf{x}$ and T, P interchangeable. We could write an expression similar to equation [5.9] for the scattering coefficient σ_{s}, using number densities and scattering cross sections for solid particles or Rayleigh scattering.

Considerable inputs go into the computation of the absorption and emission coefficients. For example, to compute $\alpha(T, P, \nu)$ for a broad range of frequencies for water vapor, up to hundreds of millions of molecular lines must be considered, due to the numerous transitions possible among the many electronic, rotational, and vibrational states of the water vapor molecules.

5.2.5 Mass-Independent Opacity

Until now we have defined the opacity as the extinction coefficient in units of m^{-1}. Many books and articles instead use the mass-independent opacity, defined as

$$\kappa_{\mathrm{m}}(T, P, \nu) = \sum_j \frac{\kappa(T, P, \nu)_j}{\rho_j}, \tag{5.10}$$

in units of m^2/kg. Here j refers to a gas species j. ρ is the gas density and for planetary atmospheres can be described by the ideal gas law.

The mass-independent opacity is useful because it is independent of the number density of molecules in the atmosphere. To further explore the utility of this opacity, let us revisit the opacity definition equation [5.9]. The opacity is equal to the number density times the absorption (or scattering) cross section. The number density itself depends heavily on pressure and temperature through the ideal gas law. The simplicity of the mass-independent opacity, therefore, is that it varies much less with pressure and temperature than does the mass-dependent opacity.

5.2.6 Mean Opacities

In many treatments of radiative transfer we desire a mean opacity so that we may consider equations averaged over frequency (i.e., wavelength). At first thought, we might consider the average or mean opacity taken literally as the average opacity value for all frequencies. In practice such an opacity average would not be very useful. Consider the following: the flux of a planet approximated as a black body (Figure 3.8); and a planet with exceedingly high opacity at UV frequencies, but close to zero opacity at all other frequencies. In this case an opacity average is not useful—the high opacity is at UV frequencies where there is virtually no planetary flux. In this artificial example, a better representation would be simply zero opacity.

The point is that in radiative transfer we are interested in how much flux is blocked by (or allowed to pass through) the atmosphere at different frequencies by the absorbing or emitting particles. Useful mean opacities are therefore those that are weighted by a function of the intensity. Because the intensity of a planet atmosphere is unknown—and indeed the quantity we are trying to solve for—the black body intensity is used.

The Planck mean opacity is defined as

$$\kappa_{\mathrm{P}}(T,P,\nu) = \frac{\int_0^\infty \kappa(T,P,\nu)B(T,\nu)d\nu}{\int_0^\infty B(T,\nu)d\nu}. \tag{5.11}$$

The Planck mean opacity is the opacity weighted by the black body intensity at a given temperature. Where the opacity is high, the contribution to the mean opacity is also high. The Planck mean opacity is valid in "optically thin" regimes (see Section 5.3 for a definition of optically thin).

The Rosseland mean opacity is defined as

$$\frac{1}{\kappa_{\mathrm{R}}(T,P,\nu)} = \frac{\int_0^\infty \frac{1}{\kappa(T,P,\nu)}\frac{dB(T,\nu)}{dT}d\nu}{\int_0^\infty \frac{dB(T,\nu)}{dT}d\nu}. \tag{5.12}$$

The Rosseland mean opacity is a harmonic mean, weighted by the temperature derivative of the black body intensity. The Rosseland mean is useful because it gives a higher weight to frequencies with a small opacity than to frequencies with a large opacity—this captures the physical situation where more radiation travels through the atmosphere at frequencies where opacity is smallest. This is in contrast to the Planck mean opacity. Typically the Planck mean opacity is valid in optically thin regions of the atmosphere and the Rosseland mean opacity is valid in optically thick regimes (see Section 5.3 for definitions of optically thin and thick). To further understand the definition of the Rosseland mean opacity, we must wait until a discussion of radiative diffusion in Section 6.4.4.

5.3 OPTICAL DEPTH

In radiative transfer we are aiming to understand the interactions of photons (or a beam of photons) as they travel through a planetary atmosphere. From our viewpoint, it is straightforward to think of a distance in meters or kilometers in a planetary atmosphere. A more natural distance scale for photon interactions is the optical depth scale. As we shall see in this chapter and the next, the optical depth scale enables a simplification of the radiative transfer equation and its solutions.

The optical depth τ is a measure of transparency and describes how opaque a part of the planetary atmosphere is to radiation traveling through it. A medium that is completely transparent has an optical depth of zero. A planetary atmosphere usually has some optical depth. In the optically thin case, $\tau \ll 1$; a photon could travel the distance s without being absorbed or scattered. The opposite is the optically thick case, $\tau > 1$. We will later see that τ is the dimensionless e-folding factor for absorption of hot radiation through a cooler gas layer. The optical depth is related to the absorbing particles in a planetary atmosphere. Because the extinction coefficient depends on frequency and on location in the planet atmosphere, so too does the optical depth. We therefore write the optical depth as $\tau(\mathbf{x}, \nu)$.

The definition of the optical depth along a one-dimensional path s is

$$\boxed{d\tau(s, \nu) = -\kappa(s, \nu) ds.} \tag{5.13}$$

The optical depth is dimensionless.

The optical depth is often used as a distance scale in planetary atmospheres. In plane-parallel atmospheres (see Section 5.6.2) the optical depth scale often takes the convention of being measured backward along the ray of traveling photons. In other words, the plane-parallel atmosphere optical depth scale has $\tau = 0$ defined at the top of the atmosphere, because the observer is looking down into the planetary atmosphere. This convention introduces a negative sign for the optical depth scale.

$$\boxed{\tau(z, \nu) = -\int_{z=0}^{z_{\max}} \kappa(z', \nu) dz' \equiv \int_{z_{\max}}^{z=0} \kappa(z', \nu) dz'.} \tag{5.14}$$

For notational simplicity we use τ_ν and take the z-dependence as implied.

We can relate the optical depth to the mean free path of a photon, l. (In this description we will drop the variable dependencies for clarity.) The mean free path is defined as the mean distance a photon can travel before it is stopped by interaction with a molecule or other atmospheric particle. Let us consider a slab with area A^2, volume $A^2 dz$, and number density n. The probability of an incident photon stopping in the slab can be calculated as the ratio of the net area of the molecules to the slab area,

$$P(z) = \frac{n\sigma A^2 dz}{A^2} = n\sigma dz, \tag{5.15}$$

where σ is the cross section. But we have already described the drop in intensity of a beam of radiation,

$$dI = -\kappa I dz = -n\sigma I dz. \tag{5.16}$$

We can now see a reasonable definition for the mean free path

$$l = \frac{1}{n\sigma} = \frac{1}{\kappa}, \tag{5.17}$$

and we can relate

$$d\tau = \frac{1}{l} dz. \tag{5.18}$$

Let us now recover the mean free path more formally by starting with $dI = -\kappa I dz$, or

$$\frac{dI}{I} = -\frac{dz}{l}, \tag{5.19}$$

with a solution

$$I = I_0 \mathrm{e}^{-z/l}. \tag{5.20}$$

Now, the probabiity that a photon is absorbed between z and $z + dz$ is

$$dP(z) = \frac{I(z) - I(z+dz)}{I_0} = -\frac{1}{l}\mathrm{e}^{-z/l}dz. \tag{5.21}$$

We take can finally use the expectation value to show that the mean free path is l,

$$\langle z \rangle = \int_0^\infty z dP(z) = \int_0^\infty \frac{z}{l}\mathrm{e}^{-z/l}dz = l. \tag{5.22}$$

5.4 LOCAL THERMODYNAMIC EQUILIBRIUM

5.4.1 LTE Definition

We now arrive at one of the most important fundamental concepts in radiative transfer. This is the concept of local thermodynamic equilibrium (LTE). Complete thermodynamic equilibrium applies when the material is in thermal, chemical, and mechanical equilibrium. On the contrary, across a planet atmosphere, we expect huge differences in temperature and pressure, especially because of the open boundary at the top of the atmosphere. LTE is valid in a local area of the atmosphere where any temperature, pressure, or chemical gradients are small compared to the photon mean free path. LTE is therefore a local version of complete thermodynamic equilibrium. More specfically, in LTE, we assume that all of the conditions in thermodynamic equilibrium hold, except we let the radiation field depart from that of a black body. This is because one of the fundamental properties of planetary atmospheres is a radiation field that is very different from the black body radiation. We must use the radiative transfer equation to compute the intensity.

The reason LTE is one of the most important fundamental concepts in radiative transfer is that it provides a sweeping simplification of the radiative transfer problem. LTE will, in fact, enable us both to understand radiative transfer and spectral line formation conceptually, and to simplify the numerical solutions to the equation of radiative transfer.

5.4.2 LTE and Level Populations

By now you may be wondering: what is the magical simplification using LTE and how does it help to solve the radiative transfer equation? Let us go back to the absorption coefficient α (equation [5.3]) and the emission coefficient ε (equation [5.4]). Recall that these coefficients are made up of a number density times an absorption (or scattering) cross section (see Section 5.2.4 and equation [5.9]). The total number density times cross sections are summed over all level populations. By level populations we mean the numbers of an atom or molecule in the different quantum mechanical energy states. In order to compute κ and ε, we need to know (1) the overall number densities of atoms or molecules and (2) the fraction of atoms or molecules in each quantum state. More formally, we would say that we need to know the state of the matter, the chemical partitioning of the matter, and the level populations.

The heart of the problem is that all of the above (in particular the level populations) are determined by the radiation field—yet the radiation field is the key variable that we are trying to solve for in the radiative transfer equation. As an example of how the radiation field and the level populations are coupled, let us take photoexcitation followed by photodeexcitation. An electron in an atom or molecule may be excited to a higher level by absorption of a photon. Later, due to either spontaneous or stimulated emission, the electron will cascade downward. This electron transition will release photons, that is, release energy as radiation. We could say this photoexcitation (and stimulated emission) is controlled by properties of the radiation field. Therefore, in this example, the level populations are controlled by the radiation field. This coupled nature of the radiation field and the level populations is a significant hurdle in solving the radiative transfer equation. In principle a full solution of the radiation field and the level populations is possible; in practice this is onerous.

LTE makes the solution of radiative transfer easier because we can assume we know how energy levels of different atoms and molecules are populated: they depend only on the local kinetic temperature and one other variable (such as ρ or P). More formally we say that LTE assumes a decoupling of the state of the matter and the radiation field by explicitly assuming that all properties of the matter depend only on the local kinetic temperature and density (or pressure). In LTE, the atomic and molecular energy level populations can be computed using the equilibrium relations of statistical mechanics (see Chapter 8). For atomic and molecular energy levels, the relevant equation is the Boltzmann distribution, equation [8.30].

5.4.3 Conditions for LTE

Under what conditions is LTE valid? The concept of LTE implies a strict connection of the matter component to the local temperature. LTE is therefore valid where this matter-temperature connection is physically realized.

The conditions relevant for planetary atmospheres are at high enough densities that collisional processes dominate over competing radiative processes. Constant collisions enable matter and radiation to share the same temperature, and hence LTE to be valid. For example, an electron in an atom or molecule may be excited to a higher level by absorption of a photon. Later, due to a collision, the electron will cascade downward, releasing energy as the colliding particles' kinetic energy. This energy eventually ends up as a part of the thermal pool, after the particles undergo further elastic collisions with other atoms or molecules in the gas. In this case of collisional deexcitation, the photon is said to be destroyed and converted into kinetic energy of the gas. In this way the collisional processes couple the photons (radiation field) to the matter temperature via the kinetic energy of the gas. (On a more subtle note, as long as the colliding particles have velocities in a Maxwellian distribution, the energy level populations will have their Boltzmann "equilibrium" values.)

The second situation in which LTE is valid is not really relevant for planetary atmospheres. This situation is where the mean radiation field is a black body. (Note that, from the above example, this is not a necessary condition of LTE). In the case that $J(\mathbf{x}, \nu) = B(\mathbf{x}, \nu)$, the radiative rate equations that control the level populations will give an equilibrium (i.e., Boltzmann) distribution for the level populations (see Section 8.2.4).

For a given situation one could try to determine the validity of LTE by computing and comparing radiative and collisional rates, but to accurately validate LTE a non-LTE calculation itself may be necessary. LTE is actually valid in Earth's atmosphere from the ground up to about 60 km. This is largely because collisional rates between molecules dominate radiative rates, driving a Boltzmann distribution. We know from spectral measurements of Earth at the top of the atmosphere and also at different altitudes that the Earth atmosphere radiation field is not a black body.

Despite the fact that LTE is commonly used in exoplanet atmosphere calculations, there are some regions of the atmosphere where LTE is not valid. One example is in the very upper layers of the planet atmosphere where radiation freely escapes through the open boundary. Here, both the radiation field is very different from a black body and collisions do not dominate over radiative transition rates, causing a departure from LTE.

5.4.4 LTE and Kirchhoff's Law

We now turn to Kirchhoff's Law, one of the LTE expressions that simplifies the radiative transfer equation. Kirchhoff's Law of radiation states that at thermal equilibrium the emissivity of a body equals its absorptivity. Kirchhoff's Law is

$$\boxed{\varepsilon_{\text{therm}}(\mathbf{x}, \nu) = \alpha(\mathbf{x}, \nu) B(\mathbf{x}, \nu).} \tag{5.23}$$

This means the thermal emission is related to the absorptive properties of the gas. The thermal emission at a given location $\mathbf{x}$ in the planetary atmosphere is black body radiation weighted by the the absorptive properties of the gas. In other words, any time a photon is absorbed in a gas in LTE, the energy will be reemitted as radiation, with the amount and wavelength depending on the Planck function weighted by the absorption coefficient.

Kirchhoff's Law, is, technically, valid only under conditions of complete thermodynamic equilibrium. In planetary atmospheres we justify using Kirchhoff's Law for small, localized areas of the atmosphere where gradients in thermodynamic properties are small compared to a photon mean free path.

5.5 THE SOURCE FUNCTION

The source function (also sometimes called the contribution function) is defined as the ratio of the emission coefficient to the extinction coefficient:

$$\boxed{S(\mathbf{x}, \hat{\mathbf{n}}, \nu) = \frac{\varepsilon(\mathbf{x}, \hat{\mathbf{n}}, \nu)}{\kappa(\mathbf{x}, \nu)}.} \tag{5.24}$$

The source function is used to simplify the radiative transfer equation and solution and to provide physical insight. The source function has the same dimensions as intensity.

We can expand the source function by breaking it up into a thermal emission component and a scattering component. We have previously used Kirchhoff's Law for thermal emission, $\varepsilon_{\text{therm}} = \alpha B$. For isotropic scattering, $\varepsilon_{\text{scat}} = \sigma_{\text{s}} J$. So, under conditions of LTE where Kirchhoff's Law is valid and where scattering is coherent and isotropic we may write the source function as

$$\boxed{S(\mathbf{x}, \nu) = \frac{\alpha(\mathbf{x}, \nu) B(\mathbf{x}, \nu) + \sigma_{\text{s}}(\mathbf{x}, \nu) J(\mathbf{x}, \nu)}{\alpha(\mathbf{x}, \nu) + \sigma_{\text{s}}(\mathbf{x}, \nu)}.} \tag{5.25}$$

This expression differs from the strict LTE value of the source function. In LTE, $S(\mathbf{x}, \nu) = B(\mathbf{x}, \nu)$, according to Kirchoff's Law, equation [5.23]. The above source function includes scattering terms and therefore allows a slight deviation from LTE.

In LTE with no scattering we can use Kirchhoff's Law to write the source function as

$$S(\mathbf{x}, \nu) = B(\mathbf{x}, \nu). \tag{5.26}$$

In general, though, while $I(\mathbf{x}, \hat{\mathbf{n}}, \nu) = B(\mathbf{x}, \nu)$ for black body radiation, LTE radiation is not necessarily black body radiation. Therefore in LTE I is not required to equal B.

5.6 THE EQUATION OF RADIATIVE TRANSFER

5.6.1 The Time-Dependent Equation

For completeness we now return to a derivation of the *time-dependent* radiative transfer equation. Here we follow both the outline at the beginning of this chapter as well as the deriviation in [1].

We first point out that, for moving material, both the extinction coefficient κ and the emission coefficient ε have a dependence on angle. This is because in moving material, changes in direction (and hence angle) result from the Doppler shift.

We choose an inertial coordiate system and want to understand the energy flow along a path s in a direction $\hat{\mathbf{n}}$ into a differential solid angle $d\Omega$. The intensity beam is traveling through a fixed volume element of length ds and cross section dA normal to $\hat{\mathbf{n}}$, in a time interval dt. As before, we add up the losses from and

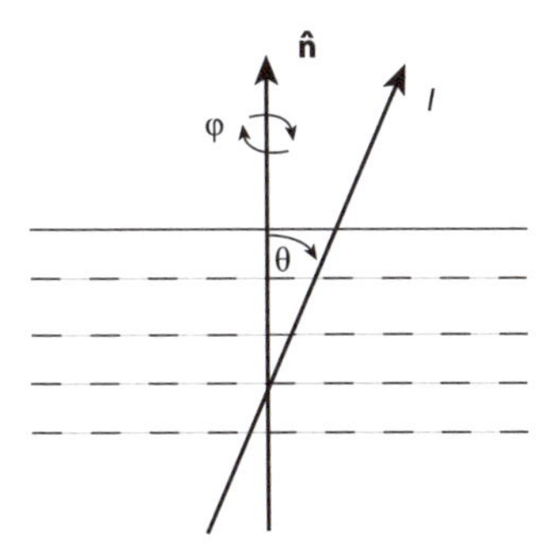

Figure 5.2 Schematic description of a 1D plane-parallel atmosphere.

gains into the beam of radiation along the pathlength of the beam, but now we also consider the losses and gains as functions of time,

$$\begin{aligned} &[I(\mathbf{x} + \Delta\mathbf{x}, \hat{\mathbf{n}}, \nu, t + \Delta t) - I(\mathbf{x}, \hat{\mathbf{n}}, \nu, t)]dAd\Omega d\nu dt \\ &= [-\kappa(\mathbf{x}, \hat{\mathbf{n}}, \nu, t)I(\mathbf{x}, \hat{\mathbf{n}}, \nu, t) + \varepsilon(\mathbf{x}, \hat{\mathbf{n}}, \nu, t)]dsdAd\Omega d\nu dt. \end{aligned} \tag{5.27}$$

The pathlength Δs and the time interval Δt are related by $\Delta t = \Delta s/c$, and so

$$\begin{aligned} &[I(\mathbf{x} + \Delta\mathbf{x}, \hat{\mathbf{n}}, \nu, t + \Delta t) - I(\mathbf{x}, \hat{\mathbf{n}}, \nu, t)] \\ &= \left[\frac{1}{c}\left(\frac{\partial I(\mathbf{x}, \hat{\mathbf{n}}, \nu, t)}{\partial t}\right) + \left(\frac{\partial I(\mathbf{x}, \hat{\mathbf{n}}, \nu, t)}{\partial s}\right)\right] ds. \end{aligned} \tag{5.28}$$

We now substitute equation [5.28] into equation [5.27] to get the time-dependent radiative transfer equation

$$\boxed{\left[\frac{1}{c}\left(\frac{\partial}{\partial t}\right) + \left(\frac{\partial}{\partial s}\right)\right] I(\mathbf{x}, \hat{\mathbf{n}}, \nu, t) = -\kappa(\mathbf{x}, \hat{\mathbf{n}}, \nu, t)I(\mathbf{x}, \hat{\mathbf{n}}, \nu, t) + \varepsilon(\mathbf{x}, \hat{\mathbf{n}}, \nu, t).} \tag{5.29}$$

For the rest of this chapter (and for some future chapters) we will discuss the static case, independent of time. We will therefore drop the time-dependent terms of the above radiative transfer equation, including the $\hat{\mathbf{n}}$-dependence of the extinction coefficient κ. As previously in this chapter, we will omit the t-dependence of the intensity, flux, and other terms. Nevertheless, we must still keep in mind that radiation involves energy flow: energy per unit time.

5.6.2 The Plane-Parallel Approximation

The plane-parallel atmosphere is a good framework in which to study the radiative transfer equation (Figure 5.2). For a 1D planar atmosphere the atmosphere is defined by a stratified plane with each layer having homogeneous properties such as T, P, and ρ. This 1D plane represents a location on the surface of the planet. In what case is the plane-parallel approximation valid? If the radial depth of the atmosphere is much smaller than the planetary radius.

The plane-parallel definition assumes axial symmetry. Considering the axial symmetry and adopting z as our (1D) vertical coordinate, we have

$$\frac{\partial}{\partial x} = \frac{\partial}{\partial y} = 0 \tag{5.30}$$

and

$$\frac{dz}{ds} = \cos\theta \equiv \mu, \tag{5.31}$$

where θ is the angle between the surface normal and a beam of intensity. With the above we therefore have the plane-parallel radiative transfer equation

$$\mu \frac{dI(z, \hat{\mathbf{n}}, \nu)}{dz} = -\kappa(z, \nu) I(z, \hat{\mathbf{n}}, \nu) + \varepsilon(z, \hat{\mathbf{n}}, \nu). \tag{5.32}$$

If we further consider axial symmetry, that is, no ϕ dependence,

$$\boxed{\mu \frac{dI(z, \mu, \nu)}{dz} = -\kappa(z, \nu) I(z, \mu, \nu) + \varepsilon(z, \mu, \nu).} \tag{5.33}$$

As written, the plane-parallel radiative transfer equation is an integro-differential equation: because $\varepsilon = \varepsilon_{\rm scat} + \varepsilon_{\rm therm}$ and the scattering term contains angle integrals of I,

$$\varepsilon_{\rm scat}(z, \mu, \nu) = \sigma_{\rm s}(z, \nu) \frac{1}{4\pi} \int_0^{2\pi} \int_{-1}^{1} P(\mu, \phi; \mu', \phi') I(z, \mu', \nu) d\mu' d\phi'. \tag{5.34}$$

In a plane-parallel atmosphere the mean intensity (equation [2.2]) and flux (equation [2.5]) can also be rewritten considering azimuthal symmetry and the definition of μ,

$$J(z, \nu) = \frac{1}{4\pi} \int_0^{2\pi} \int_{-1}^{1} I(z, \mu, \nu) d\mu d\phi, \tag{5.35}$$

or

$$\boxed{J(z, \nu) = \frac{1}{2} \int_{-1}^{1} I(z, \mu, \nu) d\mu,} \tag{5.36}$$

$$F(z, \nu) = \int_0^{2\pi} \int_{-1}^{1} \mu I(z, \mu, \nu) d\mu d\phi, \tag{5.37}$$

or

$$\boxed{F(z, \nu) = 2\pi \int_{-1}^{1} \mu I(z, \mu, \nu) d\mu.} \tag{5.38}$$

(We note that in the stellar atmosphere literature the so-called astrophysical flux $H = \frac{1}{4\pi} F$ is often used because it is similar to the form for J and K.)

In preparation for solving the radiative transfer equation we first rewrite the equation [5.33] using the optical depth distance scale (described in Section 5.3) and the definition of the source function (described in Section 5.5):

$$\boxed{\mu \frac{dI(\tau_\nu, \mu, \nu)}{d\tau_\nu} = I(\tau_\nu, \mu, \nu) - S(\tau_\nu, \mu, \nu).} \tag{5.39}$$

We emphasize the difference between incoming rays ($\mu < 0$) and outgoing rays ($\mu > 0$).

The upper boundary condition for this differential equation is the stellar radiation incident on the planet atmosphere as a function of direction μ,

$$I(0, \mu, \nu) = I_*(0, \mu_0, \nu) \qquad (-1 \leq \mu \leq 0). \tag{5.40}$$

For plane-parallel radiation, the incoming stellar radiation is incident at one angle μ_0 only (see Section 2.7.) For planets with no incident radiation, for calculations at infrared wavelengths when the planetary emission completely dominates any incoming stellar radiation, or for the "dark," nonilluminated hemisphere of a planet, $I_* = 0$.

The lower boundary condition is the intensity coming from the planet interior

$$I(\tau_{\max,\nu}, \mu, \nu) = I_{\rm int}(\tau_{\max,\nu}, \mu, \nu) \qquad (0 \leq \mu \leq 1). \tag{5.41}$$

The intensity in this lower boundary condition is matched to the planet's interior energy—the energy coming from the deep interior and incident on the lower boundary of the atmosphere.

5.6.3 The Formal Solution

A formal solution to the 1D plane-parallel radiative transfer equation (equation [5.39]) may be obtained with the integrating factor $e^{-\tau_\nu/\mu}$, whereby the radiative transfer equation is written

$$\frac{dI(\tau_\nu, \mu, \nu)}{d\tau_\nu} e^{-\tau_\nu/\mu} - I(\tau, \mu, \nu)\frac{1}{\mu}e^{-\tau_\nu/\mu} = -\frac{1}{\mu}S(\tau_\nu, \mu, \nu)e^{-\tau_\nu/\mu}. \tag{5.42}$$

Integration from an initial optical depth $\tau_{\nu,\rm i}$ to a final optical depth $\tau_{\nu,\rm f}$ gives the solution

$$I(\tau_{\nu,\rm f}, \mu, \nu) = \\ I(\tau_{\nu,\rm i}, \mu, \nu)e^{-(\tau_{\nu,\rm i} - \tau_{\nu,\rm f})/\mu} - \frac{1}{\mu}\int_{\tau_{\nu,\rm i}}^{\tau_{\nu,\rm f}} S(\tau'_\nu, \mu, \nu)e^{-(\tau'_\nu - \tau_{\nu,\rm f})/\mu} d\tau'_\nu. \tag{5.43}$$

The solution has two terms on the right side of the equation. The first term describes the initial intensity diminished by exponential attenuation of absorption. The second term describes the emission from the atmosphere: an exponentially weighted average of the source function along the beam up to the location of interest.

Our goal in solving the radiative transfer equation is to derive a planet spectrum: the emergent flux at the top of the atmosphere. The emergent flux is the measurable quantity for exoplanetary atmospheres. For a semi-infinite atmosphere, integration from deep in the planetary atmosphere $\tau_{\nu,\rm i} = \infty$ to the top of the planet atmosphere at $\tau_{\nu,\rm f} = 0$, the emergent intensity is

$$I(0, \mu, \nu) = \frac{1}{\mu}\int_0^\infty S(\tau_\nu, \mu, \nu)e^{-\tau_\nu/\mu} d\tau_\nu, \tag{5.44}$$

where we have used

$$\lim_{\tau_\nu \to \infty} I(\tau_\nu, \mu, \nu)e^{-\tau_\nu/\mu} = 0. \tag{5.45}$$

The emergent intensity is the amount of intensity at each altitude that reaches the surface along a path with angle θ to the line of sight.

The emergent surface flux can be derived from equation [2.10],

$$\boxed{F(0,\nu,t) = 2\pi \int_0^1 \int_0^\infty S(\tau_\nu,\mu,\nu)\mathrm{e}^{-\tau_\nu/\mu} d\tau_\nu d\mu.} \tag{5.46}$$

If the source function $S(\tau,\nu,\mu)$ is known, the emergent intensity and the emergent flux (i.e., the planet's spectrum) can be computed directly from the above two equations. In many cases, however, this straightforward solution is not possible. The main complication is that the source function itself depends on $I(\tau,\mu,\nu)$, the quantity we are trying to solve for.

As a specific example let us consider that the source function depends on the intensity through any scattering terms that may also be μ-dependent. Recall that the source function may be described as

$$S(\tau,\nu,\mu) = \frac{\varepsilon(\tau,\nu,\mu)}{\kappa(\tau,\nu)}, \tag{5.47}$$

where the emission coefficient ε has an angle-independent thermal emission component $\varepsilon_{\rm therm}(\tau_\nu,\nu) = \alpha B(\tau_\nu,\nu)$ and an angle dependent scattering component of ε (equation [5.34]),

$$\varepsilon_{\rm scat}(\tau,\mu,\nu) = \sigma(\tau,\nu)\frac{1}{4\pi}\int_0^{2\pi}\int_{-1}^{1} P(\mu,\phi;\mu',\phi')I(\tau,\mu',\nu)d\mu' d\phi'. \tag{5.48}$$

Physically the scattering term (whether angle dependent or isotropic) means that the intensity is decoupled from local conditions: the photons may scatter through large distances in the atmosphere without interacting with the thermal pool of the gas via absorption and thermal reemission. The scattering term requires a numerical solution of an integro-differential equation.

To further investigate the hidden complication in solving the 1D plane-parallel radiative transfer equation [5.39] we return to a discussion of the boundary conditions, equations [5.40] and [5.41]. To solve for intensity from this first-order ordinary differential equation we require two full boundary conditions, that is, an upper and a lower boundary condition on the full range $(-1 \leq \mu \leq 1)$. Yet, the information we have is the stellar radiation incident on the planet and traveling downward $(-1 \leq \mu < 0)$, and an estimate of the interior energy that we may convert to an outward-going intensity $(0 < \mu \leq 1)$ at the lower boundary. Because the boundary conditions are not fully specified, iterative techniques or a different formulation of the radiative transfer equation are needed to solve for I.

5.7 SUMMARY

We have presented fundamental concepts leading up to the foundational equation of radiative transfer. We started with opacity, the macroscopic description that captures the interaction of radiation with gases or solids in the atmosphere. Opacity is a major component of the radiative transfer equation. The concept of optical depth is related to opacity; an opaque atmosphere is optically thick and a transparent atmosphere is optically thin. The quantitative optical depth scale is a useful substitution for the distance scale in solving the radiative transfer equation. We defined the source function, a convenient ratio of emission to extinction for solving the radiation transfer equation. The concept of local thermodynamic equilibrium (LTE) was described in some detail. The situation of LTE enables a decoupling of radiation from the local temperature, which greatly simplifies the radiative transfer problem and solution; the atomic and molecular energy population levels are specified by the local temperature and do not have to be determined by a simultaneous solution with the radiative transfer equation. We finally came to the radiative transfer equation itself, an equation that can be set out in 1D as the rate of change of intensity (a beam of traveling photons) with distance is equal to the loss from the beam and the additions to the beam. We are now ready to proceed to solutions of the radiative transfer equation.

REFERENCES

For further reading

For the most thorough description of LTE in atmospheres:

• Mihalas D. 1978. *Stellar Atmospheres* (2nd ed.; San Francisco: W. H. Freeman).

For a concise outline of the radiative transfer fundamental concepts:

• Chapter 1 in Rybicki, G. B., and Lightman, A. P. 1986. *Radiative Processes in Astrophysics* (New York: J. Wiley and Sons).

See Chapter 6 for a more complete annotated list of radiative transfer textbooks.

EXERCISE

1. Kirchhoff's Law of radiation. Explain why Kirchhoff's Law is valid in thermodynamic equilibrium. Use a conceptual explanation, based on absorption and emission in a black body enclosure. Also use the 1D plane-parallel equation of radiative transfer.

Chapter Six

Radiative Transfer II: Solutions

6.1 INTRODUCTION

Radiative transfer describes how radiation changes as it travels through a medium. Although we can only observe the radiation emerging from the planet's atmosphere, much information is contained in the emergent spectrum, including details of the atmospheric temperature, pressure, and composition. For this chapter we will work under the approximations of a 1D, time-independent, homogeneous atmosphere in hydrostatic equilibrium with a known vertical temperature-pressure structure. Relaxation of some of these assumptions will be dealt with in later chapters.

The radiative transfer equation describes the change in a beam of radiation as it travels some distance s through a volume of gas. We derived the equation of radiative transfer in Chapter 5. For review, we recall the plane-parallel form of the equation of radiative transfer and its origin. The changes in a beam of traveling photons are due to losses from the beam $\kappa(z,\nu,t)I(z,\mu,\nu,t)$, where κ is the extinction coefficient, and additions to the beam $\varepsilon(z,\mu,\nu,t)$, where ε is the emission coefficient. For a plane-parallel atmosphere we use the definition $dz/ds = \cos\theta \equiv \mu$. By addition of the losses and gains, for a time-independent situation, the plane-parallel radiative transfer equation is (equation [5.33]),

$$\boxed{\mu\frac{dI(z,\mu,\nu)}{dz} = -\kappa(z,\nu)I(z,\mu,\nu) + \varepsilon(z,\mu,\nu).} \qquad (6.1)$$

Because we are considering time-independent equations, in this chapter we will omit the dependency t.

The goal of this chapter is to gain an understanding of the emergent planetary spectrum. We begin with a conceptual understanding of line formation in a planetary spectrum. We then explore approximate solutions to the radiative transfer equation to connect the emergent spectrum with conditions in the exoplanetary atmosphere. We emphasize physics and a conceptual and analytical understanding rather than a description of computational approaches.

6.2 A CONCEPTUAL DESCRIPTION OF THE EMERGENT SPECTRUM

At its heart a planetary spectrum has two peaks: a visible wavelength peak due to scattered starlight and an infrared peak due to the thermal emission from the planet atmosphere. An exoplanet's thermal emission has two possible origins. The first is from reprocessed absorbed stellar radiation and the second is from interior energy

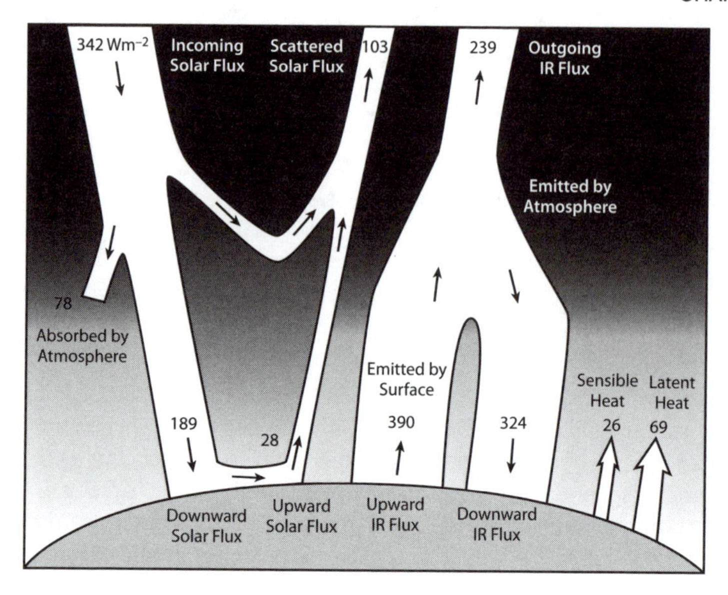

Figure 6.1 Energy balance of Earth's atmosphere.

leaking out through the atmosphere (arising from slow loss of residual gravitational potential energy and from radioactive decay). The relative fraction of reprocessed absorbed stellar energy versus interior energy is different for each exoplanet. Hot Jupiters, for example, are so close to their stars that their thermal emission from reprocessed absorbed stellar radiation overwhelms any contribution from interior energy. Figure 3.8 shows the relative fluxes for the Sun, Jupiter, Earth, Venus, Mars, and a putative hot Jupiter, approximating each as a black body, illustrating the visible and the thermal infrared flux peaks. Earth's atmospheric processes are illustrated in terms of energy balance in Figure 6.1. Many of these atmospheric processes are imprinted on Earth's spectrum.

Spectral lines hold the key to the planetary atmosphere conditions: the chemical composition and the vertical temperature and pressure distribution. A solution to the radiative transfer equation gives the full exoplanet atmosphere emergent spectrum. In this section our aim is to gain a conceptual understanding of spectral line formation, before attempting to explore solutions to the radiative transfer equation.

6.2.1 General Foundation

We begin by reviewing the conditions that can generate three different kinds of spectra: a continuous spectrum; a spectrum with absorption lines; and a spectrum with emission lines (see Figures 6.2 and 6.3).

We will start by considering a black body radiator. By definition, a black body emits all radiation incident on it, and this leads to a continuous spectrum (Figure 2.7). Recall that a black body continuous spectrum has an intensity and frequency distribution depending only on its temperature, given in equation [2.34].

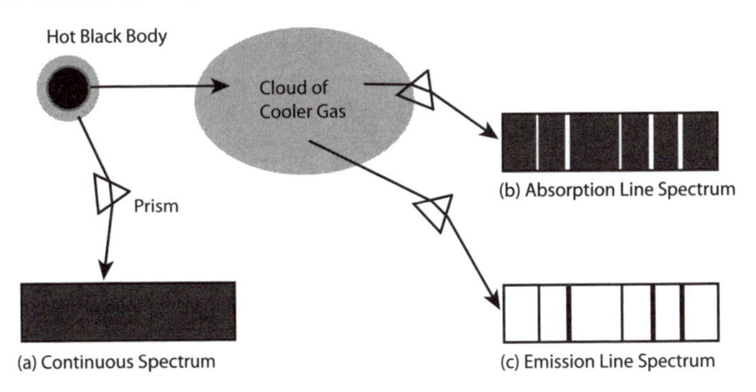

Figure 6.2 Origin of spectral lines. See text for a discussion.

If a planet atmosphere were a black body radiator, we would know its spectrum: continuous, with no spectral absorption or emission lines.

We now imagine viewing a cloud of cooler gas in front of a hotter black body radiator. We would see the continuous spectrum as emitted by the black body radiator—but the continous spectrum would be broken up by foreground absorption lines: an absorption line spectrum. The absorption lines originate when some of the black body radiation is absorbed as it travels through the cloud of cooler gas. The wavelengths of absorption depend on which atoms and molecules are present in the gas atmosphere. A colloquial way of thinking of this is as a black body continuous spectrum with "bites" taken out of it by the cooler foreground gas. What happens to the absorbed radiation? In a low-density gas, the radiation will be reemitted—isotropically. Along the line of sight, we will still see absorption features because only a small fraction of the absorbed radiation is reemitted along the line-of-sight.

We again imagine a cloud of cooler gas in front of a hotter black body radiator, but now at a viewing angle such that the black body radiator is no longer directly behind the cloud. From this viewing angle away from the black body backdrop, we would see an emission line spectrum. Where do the emission lines come from? We explained above that the atoms and molecules that absorb light emitted by the background black body radiator will eventually reemit the absorbed radiation isotropically. Some of the radiation will therefore be emitted along a line of sight for the adopted viewing angle. Without any background black body radiation, all we would see is emission. More specifically, imagine an atom in the cool gas cloud that absorbs a photon. This photon can excite an electron to a higher energy level. The excited electron will eventually cascade back down to a lower energy level, emitting a photon. The photons emitted from the downward-cascading electrons travel to us to make up an emission line spectrum.

6.2.2 The Thermal Spectrum: Absorption and Emission Line Spectrum

We are now ready to think of absorption and emission lines in a simplified atmosphere. The simplified atmosphere is one with concentric, homogeneous shells. For now we will adopt the idea that each layer is homogeneous in composition and pres-

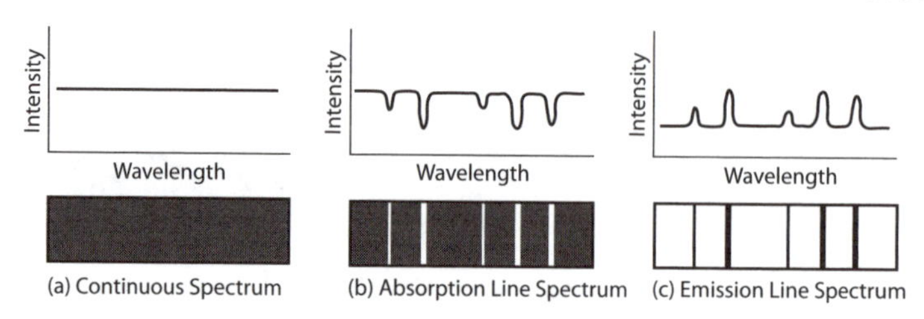

Figure 6.3 Schematic illustration of a continuous spectrum (left), an absorption line spectrum (center), and an emission line spectrum (right). This illustration is a snapshot of a narrow wavelength range so that the continuous spectrum (and the continuum for the absorption and emission line spectra) is a straight line rather than a black body curve.

sure, but that each layer differs in temperature. Furthermore, for simplicity we will describe a plane-parallel atmosphere with several atmosphere layers (Figure 6.4).

First we consider a planet with an atmosphere that is hot in the deepest layers and progressively cooler in the outer layers (the middle panel in Figure 6.4). In this scenario, we can think of each layer as a cloud of cool gas lying above a hotter gas layer. Furthermore, we assume that the deepest layer is a black body radiator. This situation is analogous to the single cool cloud in front of a black body described above. We will therefore see an absorption line spectrum.

There is an important difference from the single cool gas cloud description. While the deepest atmosphere layer may be considered to be a black body radiator, each overlying layer is not strictly a black body but can be thought of as a black body spectrum (of the layer temperature) with portions of the black body spectrum removed due to absorption in that layer. The final absorption spectrum emerging from the top of the atmosphere will have absorption from each layer.

What is the emergent spectrum in the case of a "temperature inversion," where the planet atmosphere is coolest in the innermost layer and progressively hotter with each overlying layer? By similar reasoning to the above, we would see an emission spectrum emerging from the top of the planet atmosphere.

Figure 6.4 A simplified five-layer atmosphere. Darker colors correspond to higher temperatures. See text for discussion.

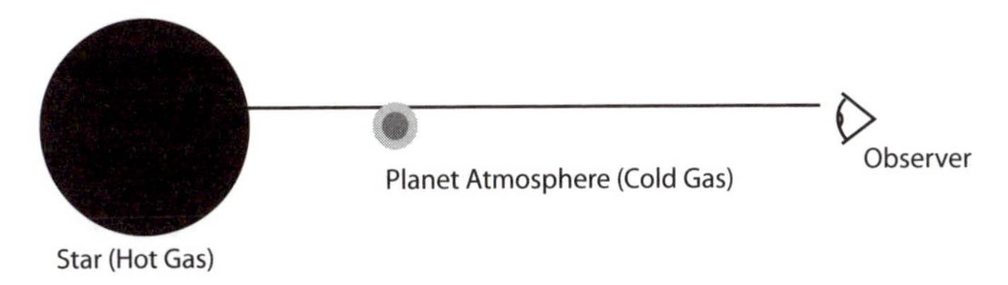

Figure 6.5 Schematic illustration of the origin of a transit transmission spectrum. The planet passes in front of the star as seen from the observer. The hot stellar radiation passes through the cooler planetary atmosphere, generating an absorption line spectrum. For illustration purposes only a single ray is shown.

We now turn to the spectrum in the idealized case of an isothermal atmosphere. If there are no cooler layers of overlying gas, there will be no absorption lines. If there are no hotter layers of overlying gas, there will be no emission lines. In our simplified picture, an isothermal atmosphere will have a continuous spectrum with no absorption or emission lines. This continuous spectrum will be a black body spectrum at the temperature of the isothermal atmosphere.

6.2.3 The Transmission Spectrum

For exoplanets, a transmission spectrum refers to a spectrum created by hot radiation passing through a cooler gas. In this case, the hot radiation is from the star and the cooler gas is the exoplanetary atmosphere. This situation, schematically represented in the top right of Figure 6.2, is shown for an exoplanet in Figure 6.5.

When the hotter stellar radiation passes through the cooler planetary atmosphere, some of the stellar radiation is absorbed by molecules and atoms in the exoplanetary atmosphere. The planetary transmission spectrum can be isolated by dividing the spectrum of the planet and star by a spectrum of the star alone (as explained further in Section 3.5.3).

6.2.4 The Scattered Light Spectrum

A planet's scattered light spectrum is the emergent spectrum of an exoplanet that originates from the scattered stellar radiation. This is primarily visible to near-infrared radiation; the wavelength range blueward of where thermal radiation dominates (Figure 3.8). Note that we prefer to use the term "scattered" over the term "reflected." Although, for atmospheres, reflection can be defined as a change in direction of a wavefront, reflection sometimes has the connotation of a wavefront bouncing off an object. Scattering includes refraction inside a particle, diffraction around a particle, and Rayleigh scattering, which is a quantum mechanical effect.

There is no simple description of scattered stellar radiation from an exoplanet, because the scattering process itself is complex and anisotropic. In addition, clouds are three dimensional and their extent and particle size are difficult to predict with models. In a cloudless atmosphere, we can, however, approximately describe the emergent radiation. We are using "scattered light" to refer to light that is scattered in the atmosphere or from the ground and emerges at the top of the atmosphere.

At visible wavelengths, at red enough wavelengths where the importance of Rayleigh scattering drops off, we may use the concept of the transmission spectrum (Figure 6.2, top right) to approximate an exoplanet's scattered light spectrum. Let us consider the Earth's scattered light as schematically illustrated on the left-hand side of Figure 6.1. Light from the Sun in the form of visible-wavelength radiation travels through Earth's atmosphere. Like a hot gas behind a cooler cloud, some of the Sun's radiation is absorbed by molecules in the Earth's atmosphere. Of the radiation that reaches the ground, some is scattered by the ground and travels through the atmosphere, back out to space. While passing back through Earth's atmosphere, parts of this radiation are further absorbed by molecules in the atmosphere. Some of the radiation from the Sun never reaches the ground, but is scattered back upward by clouds. This radiation effectively travels through a smaller part of the atmosphere than the radiation that reaches and is scattered from the ground layer.

To first order then, we may approximate the Earth's visible-wavelength radiation by twofold absorption through Earth's atmosphere. We may take a weighted average of such two-fold absorption through the atmosphere for radiation that reaches the ground and radiation that reaches cloud layers at a specified altitude in the atmosphere.

6.3 AN INTRODUCTION TO LINE FORMATION

We now switch from a conceptual introduction to emergent spectra to an approximate analytical description. We focus on line formation in thermal spectra, that is, radiation that is emitted by the atmosphere and not transmitted through or scattered in the atmosphere.

6.3.1 The Isothermal Atmosphere

We begin our mathematical investigation of the formation of spectral lines by considering an isothermal atmosphere in LTE. We start by assuming that the deepest layer of the atmosphere is optically thick. In an optically thick layer in LTE radiation behaves like a black body and we have $I = S = B$. See Section 2.8 for a description of a black body radiator. In section 6.4.4 we will see why $I = B$ at high optical depth.

Let us choose an n-layer atmosphere with n as the deepest layer. In layer n, the radiative transfer equation is

$$\mu \frac{dI_n(z,\mu,\nu)}{dz} = \varepsilon_n(z,\mu,\nu) - \kappa_n(z,\nu) I_n(z,\mu,\nu). \tag{6.2}$$

Recall that in an optically thick layer n we have $I_n(z,\mu,\nu) = B_n(z,\nu)$. We also use Kirchhoff's Law for a localized area in LTE (equation [5.23]),

$$\varepsilon_n(z,\nu) = \kappa_n(z,\nu) B_n(z,\nu), \tag{6.3}$$

to find

$$\mu \frac{dI_n(z, \mu, \nu)}{dz} = 0. \tag{6.4}$$

In other words, the intensity in layer n does not change.

Moving up to the layer $n - 1$, the unchanging intensity derived in the above equation implies that

$$I_{n-1} = I_n = B_n. \tag{6.5}$$

The radiative transfer equation in layer $n - 1$ is

$$\mu \frac{dI_{n-1}(z, \mu, \nu)}{dz} = \varepsilon_{n-1}(z, \nu) - \kappa_{n-1}(z, \nu) B_n(z, \nu). \tag{6.6}$$

Now, because the atmosphere is isothermal, we may write $B_n = B_{n-1}$. Again using Kirchhoff's Law in layer $n - 1$ we find

$$\mu \frac{dI_{n-1}(z, \mu, \nu)}{dz} = 0. \tag{6.7}$$

Repeating this exercise for all layers overlying $n - 1$,

$$I_0 = \cdots = I_{n-2} = I_{n-1} = I_n = B_n. \tag{6.8}$$

In an isothermal LTE atmosphere the emergent intensity $I_{n=0}$ will be that of a black body with the characteristic temperature of the isothermal atmosphere. Although radiation is being absorbed and reemitted in each layer, there are no net spectral features because absorption and reemission are occurring at the same temperature and therefore at the same wavelengths.

We now repeat the investigation of spectral line formation in an isothermal LTE atmosphere, but this time by way of a solution to the radiative transfer equation, rather than with the radiative transfer equation itself. Specifically, the emergent intensity solution to the radiative transfer equation for emitting atmospheres with no scattering and with $\tau_{\rm f} = 0$ (from equation [5.44]) is

$$I(0, \mu, \nu) = \int_0^\infty \frac{1}{\mu} B(\tau_\nu, \nu) \mathrm{e}^{-\tau_\nu/\mu} d\tau_\nu. \tag{6.9}$$

Recall that τ_ν is the optical depth, and here we have used the optical depth scale as a proxy for an altitude scale. We can rewrite the black body function as $B(T(\tau_\nu), \nu)$, considering that there is a temperature associated with each optical depth (i.e., altitude). In the isothermal atmosphere, the temperature is the same everywhere and so the black body radiation is also the same everywhere. We may therefore take $B(T(\tau_\nu), \nu)$ out of the integrand of the above equation to find

$$I(0, \mu, \nu) = B(T(\tau_\nu), \nu) \int_0^\infty \frac{1}{\mu} \mathrm{e}^{-\tau_\nu/\mu} d\tau_\nu. \tag{6.10}$$

Furthermore,

$$\int_0^\infty \frac{1}{\mu} \mathrm{e}^{-\tau_\nu/\mu} d\tau_\nu = 1, \tag{6.11}$$

which gives

$$I(0, \mu, \nu) = B(T, \nu), \tag{6.12}$$

and from equation [2.13],

$$F(0,\nu) = \pi B(T,\nu). \tag{6.13}$$

We have just shown that an isothermal, LTE atmosphere emits black body radiation. A black body emits a continuous spectrum (Figure 2.7) and so the emergent spectrum from an isothermal LTE atmosphere contains no spectral lines.

6.3.2 The Effect of Temperature Gradient on Line Formation

We now move from spectral line formation in an isothermal atmosphere to line formation in an atmosphere with a temperature gradient. We begin with the case of a decreasing temperature outward. This case can be thought of as a deep layer of hot gas overlain by layers of cooler and cooler gas. We have already qualitatively described this situation as producing an absorption spectrum (Figure 6.4).

In order to investigate spectral line formation under a temperature gradient, we will compare the intensity at a wavelength in a line core, $I(\tau_\nu, \mu, \nu_l)$, to the intensity at a different wavelength, in the neighboring continuum, $I(\tau_\nu, \mu, \nu_c)$. The line core, by definition, has a higher opacity than the neighboring continuum. We therefore have $\kappa(\nu_l) > \kappa(\nu_c)$. Now consider an optical depth τ_ν. From the definition of optical depth $d\tau_\nu = \kappa(\nu)dz$, we have $z_l < z_c$. In other words, for the same value of τ_ν, the line core forms at a much deeper altitude in the planet atmosphere than the continuum. From our assumption of decreasing temperature outward we have $T_l < T_c$ and hence $B(T_l) < B(T_c)$. Finally, from the emergent intensity equation [6.9], we see that $I(\tau_\nu, \mu, \nu_l) < I(\tau_\nu, \mu, \nu_l)$: a lower intensity in the line core compared to the line wing is a spectral absorption line.

In a similar description, one can show that outward increasing temperature, a large $\kappa(\nu)$, always leads to an emission line spectrum, not an absorption line spectrum.

6.3.3 The Maximum and Minimum Flux in an Absorption Line or Band

In order to more thoroughly understand a thermal absorption line spectrum, we ask: what is the minimum flux in an absorption line? We first pause to explain that a molecular absorption band is made up of a series of many closely spaced absorption lines (Chapter 8). Hereafter we will use the term absorption band instead of absorption line. We consider the absorption band in question for a planet atmosphere with temperature decreasing outward towards the upper atmosphere layers.

An excellent illustration of the minimum and maximum flux is Earth's midinfrared spectrum in Figure 3.4. Relevant here is the strong CO_2 absorption feature centered near 15 μm. The lower dashed line, showing a black body continuous spectrum with T = 215 K, nicely demarcates the minimum flux in Earth's midinfrared spectrum. Our question can also be asked: why is the minimum flux in the CO_2 absorption band not zero?

Conceptually, the minimum flux is not zero because there is always some radiation, no matter how little, emerging at any given wavelength from a planetary atmosphere. In an LTE atmosphere, the magnitude of the emergent flux depends on the temperature at the altitude where the flux originates. In the case under discussion of temperature decreasing outward, consider a wavelength with a very low opacity. The emergent flux will come from deep in the atmosphere because with a low opacity the radiation can travel essentially unimpeded. In this example, the temperature is high at the bottom of the atmosphere, and hence the maximum amount of flux will emerge at wavelengths with low opacity. In contrast, if the opacity is very high, the flux will be reprocessed in each atmospheric layer as it travels outward; the flux that finally emerges at the top of the atmosphere must originate at an altitude near the top of the atmosphere. With colder temperatures at the top of the atmosphere, a low flux will emerge at wavelengths with high opacity. Because opacity is wavelength dependent, the altitude-varying origin of flux gives rise to a spectrum that samples different temperatures, and in the LTE case the flux has different magnitudes.

6.3.4 Spectral Lines and the Vertical Temperature-Pressure Structure

We have described how different parts of a spectral line form at different altitudes in a planetary atmosphere. Specifically, the line center forms near the top of the atmosphere and the line wings form deeper in the atmosphere. We have also seen under what conditions emission versus absorption lines form in a planetary spectrum. With a temperature decreasing toward the top of the atmosphere we expect to find an absorption line spectrum, and, conversely, with a temperature increasing toward the top of the atmosphere we expect to find an emission line spectrum.

Armed with this knowledge, we may use thermal infrared observations of spectral lines to probe a planet's temperature structure with altitude. We will use Earth as an example. Figure 1.2 shows Earth's midinfrared spectrum. Figure 9.2 shows Earth's temperature structure typical of midlatitudes. From the deep O_3 and CO_2 absorption features, we can infer that Earth's atmospheric temperature is decreasing with increasing altitude in the relatively dense part of the atmosphere where these lines are formed. Indeed, mountain climbers know that as they climb higher above Earth's surface, the temperature drops.

Let us turn to look at Earth's temperature profile with altitude. The lowest layer of Earth's atmosphere, the troposphere, is where most of the atmosphere's mass is. This is the layer where weather happens and also where the bulk of visible-to-infrared spectral features are formed. In this lowest layer we do see that the temperature is decreasing with increasing altitude above the surface, consistent with our inference from Earth's spectral absorption features.

Let us take a closer look at Earth's O_3 and CO_2 absorption bands. In the center of the absorption band, there is a narrow emission feature. This emission feature tells us that somewhere in Earth's atmosphere there is a temperature inversion, that is, the temperature is *increasing* with increasing altitude. Indeed, the second lowest layer in Earth's atmosphere, the stratosphere, has just this temperature inversion (Figure 9.2). How could we infer that the temperature inversion layer is at a higher altitude than the layer in Earth's atmosphere where the bulk of the spectral bands are forming? We will see in Chapter 9 that the atmospheric density decreases exponentially with increasing altitude. Recall, furthermore, that the opacity and hence

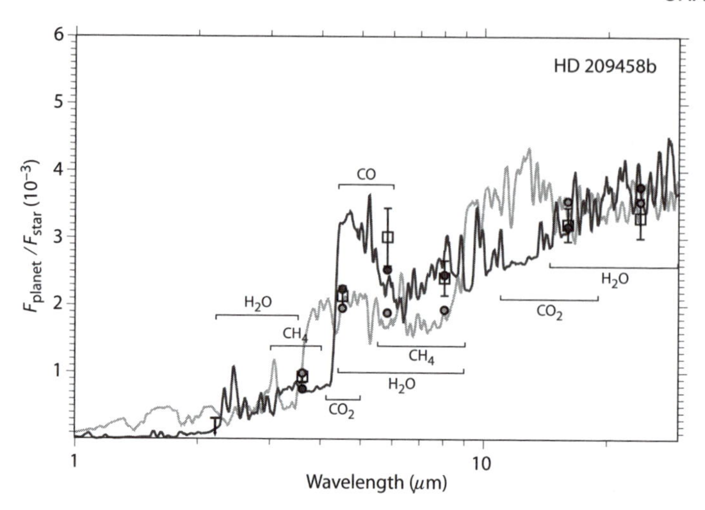

Figure 6.6 Evidence for a thermal inversion in the atmosphere of HD 209458b, assuming the presence of H_2O and CO. The squares with error bars are data points from the *Spitzer Space Telescope* [1]. The gray curve is a model with absorption features only that corresponds to an atmosphere where the temperature decreases with increasing altitude. The black curve fits the data much better and is a model with emission features only, corresponding to an atmosphere with a vertical temperature inversion. The models are from [2].

molecular band strength depends on the number density of the absorbing particle (equation [5.9]). The narrowness of the central emission feature tells us that the emission features are forming in regions of low density—much lower density than the layer where the bulk of the absorption feature is forming. This tells us that the emission band, and therefore the thermal inversion, is originating in a less dense, higher layer in the atmosphere than the layer where the bulk of the absorption band is forming.

Astronomers are able to infer temperature inversions on a subset of exoplanets. The hot Jupiters are tidally locked to their parent stars, showing the same face to the star at all times. This permanent day side can be observed by secondary eclipse observations (introduced in Section 1.2 and Figure 1.3). Data taken with the *Spitzer Space Telescope* suggest that some planets have water vapor emission features (Figure 6.6), while other hot Jupiters have the same water vapor features in absorption. Even crude spectral models can be used to infer that the emission features correspond to a planet with, on average, a deep thermal inversion on the planetary day side, as follows. Above we described Earth's spectrum where the weak emission features inside deep absorption bands indicate a temperature inversion only in a layer above where the absorption features are formed (Figure 9.2). The infrared photometry of the exoplanet HD 209458b indicates very strong emission features, with no discernible absorption bands. This suggests a temperature inversion extending deep into the atmosphere—down to the layers where the bulk of the spectral features are formed.

6.4 APPROXIMATE SOLUTIONS TO THE PLANE-PARALLEL RADIATIVE TRANSFER EQUATION

The solution of the radiative transfer equation will give us the emergent planetary spectrum. We focus our efforts on approximate solutions, with the aim of understanding the origin of the emergent spectrum. In many cases these approximations are also sufficient for basic computer codes. We remind you that throughout this chapter we are assuming that the 1D temperature-pressure vertical structure of the planet atmosphere is known (see Chapter 9 for a description of temperature-pressure profiles).

6.4.1 Transmission

A major application of transmission spectra is in transmission through an exoplanet atmosphere. During a planet transit, some of the starlight passes through the optically thin part of the planet atmosphere, picking up spectral features from the planet. A planetary transmission spectrum can be obtained by dividing the spectrum of the star and planet during transit by the spectrum of the star alone (i.e., before or after transit).

In the case of transmission of radiation alone, by definition there is no emission or scattering of photons into the beam of radiation. Therefore,

$$\varepsilon(z, \mu, \nu) = 0 \tag{6.14}$$

and hence

$$S(z, \mu, \nu) = \varepsilon(\tau_\nu, \mu, \nu)/\kappa(\tau_\nu, \nu) = 0. \tag{6.15}$$

With this simplification, the radiative transfer equation (equation [6.1]) is

$$\mu \frac{dI(z, \mu, \nu)}{dz} = -\kappa(z, \nu) I(z, \mu, \nu). \tag{6.16}$$

The radiative transfer solution from an initial position $z = z_\mathrm{i}$ to a final position $z = z_\mathrm{f}$ is

$$I(z_\mathrm{f}, \mu, \nu) = I(z_\mathrm{i}, \mu, \nu) \mathrm{e}^{-\int_{z_\mathrm{i}}^{z_\mathrm{f}} \frac{1}{\mu} \kappa(z', \nu) dz'}. \tag{6.17}$$

Using the definition of optical depth, and integrating from $\tau'_\nu = 0$ to $\tau'_\nu = \tau_\nu$, we obtain

$$\boxed{I(\tau_\nu, \mu, \nu) = I(0, \mu, \nu) \mathrm{e}^{-\tau_\nu/\mu}.} \tag{6.18}$$

The emergent intensity is just the exponentially attenuated initial intensity. This equation is known as Lambert's Law or Beer's Law; we refer to this equation as the transmitted intensity.

We will now look at an approximate solution to the transmission spectrum. Let us begin with some background information. We previously derived an expression for the planet-star flux ratio for transmission spectra, equation [3.54],

$$\frac{\mathcal{F}_{\mathrm{in\ trans},\oplus}(\nu)}{\mathcal{F}_{*,\oplus}(\nu)} = 1 - \left(\frac{R_\mathrm{p}}{R_*}\right)^2 + \frac{2 R_\mathrm{p} A_H}{R_*^2} \left(1 - \frac{\mathcal{F}_{\mathrm{S,p,trans}}(\nu)}{\mathcal{F}_{\mathrm{S},*,\mathrm{trans}}(\nu)}\right). \tag{6.19}$$

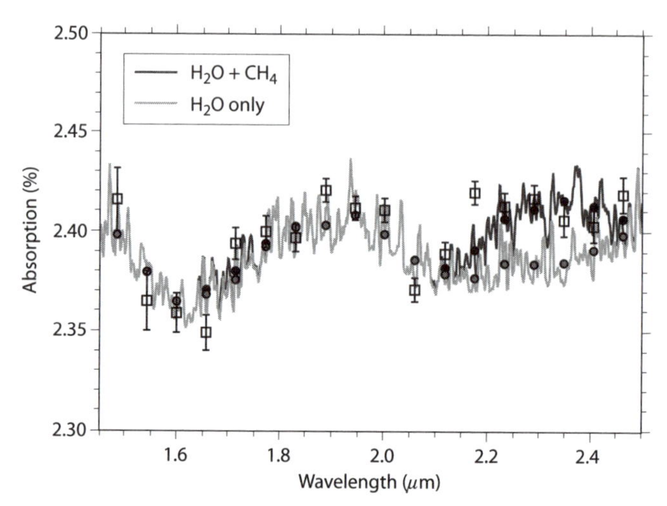

Figure 6.7 *Hubble Space Telescope* transmission spectro-photometry of the planet HD 189733 b. The data (from [3]) are shown by open squares and the models (from [2]) are shown by the solid lines. Note that the grey and black lines are coincident at the shorter wavelengths.

Here we are concerned with further understanding of the last term:

$$\frac{\mathcal{F}_{\mathrm{S,p,trans}}(\nu)}{\mathcal{F}_{\mathrm{S,*,trans}}(\nu)}, \tag{6.20}$$

where we recall that $\mathcal{F}_{\mathrm{S}}$ is the planet or star surface flux observed at Earth.

For plane-parallel radiation from the star passing through a planetary atmosphere in transmission, only the intensity at a single angle $\mu = 1$ is relevant. Our term of interest becomes

$$\frac{\mathcal{F}_{\mathrm{S,p,trans}}(\nu)}{\mathcal{F}_{\mathrm{S,*,trans}}(\nu)} = \mathrm{e}^{-\tau_\nu}, \tag{6.21}$$

because the stellar intensity cancels out in the ratio. The above equation is for one ray along the line of sight from the star to the observer (see Figure 6.5). To compute the transmission spectrum, we would have to sum up rays passing through different altitudes of the planetary atmosphere. Line broadening plays a role at higher pressures, making this consideration of rays passing through different altitudes very important. Ultimately the last term on the right-hand side of equation [6.19] must be replaced by a related more accurate term.

The list of transmission spectra of exoplanets is growing. Detected atoms and molecules for some hot Jupiters include Na, H_2O, CH_4, and CO_2. In Figure 6.7 we show data from one example, the *Hubble Space Telescope* observations of HD 189733b. The transmission spectrum signal is about 5×10^{-4}. In Chapter 3 we showed that an order-of-magnitude estimate of the transmission spectrum signal is the ratio of the planet atmosphere annulus (at a height of $5 \times H$ where H is the planetary scale height) to the area of the star. The measurements are consistent with this estimate.

6.4.2 Thermal Emission

The thermal emission approximation we are about to describe has its major application in infrared spectra of exoplanets. These include hot Jupiters and directly imaged planets—in fact any planets at wavelengths where the spectrum is due to thermal emission and scattered radiation is not important. In many ways the solution to the radiative transfer equation described in this subsection is the most powerful tool we introduce, because it is simple yet accurately solves the radiative transfer equation in the case of no scattering and in LTE. As in other solutions to the radiative transfer equation described in this chapter, we are assuming knowledge of the vertical temperature structure.

Thermal emission is the transfer of heat by radiation. In the context of planetary atmospheres, we use thermal emission to describe emergent radiation coming from the planet that is not due to scattering. Thermal radiation originates from absorbed stellar radiation, or from radiation from the planetary interior. As the radiation travels outward to the planet surface, the radiation is absorbed by atoms and molecules, heats the surrounding area, and is reemitted. Thermal emission occurs predominantly at wavelengths near peak of the black body radiation curve. For exoplanets this is typically at infrared wavelengths, though the wavelength ranges from red ($\sim$0.8 μm) for planets at $\sim$2000 K down to wavelengths of 15 μm for Earth and even longer wavelengths for colder planets (see Figure 2.7).

In this subsection we consider the case with no scattering of radiation, only with absorption and remission. This simplification is valid for infrared wavelengths where molecular absorption (and emission) is strong and dominates over scattering.

6.4.2.1 *Formal Solution*

We now turn to the radiative transfer equation and solution for thermal emission of radiation, with no scattering of photons into or out of the beam of radiation. The emission coefficient has only a thermal component,

$$\varepsilon(\tau_\nu, \mu, \nu) = \varepsilon_{\text{therm}}(\tau_\nu, \mu, \nu). \tag{6.22}$$

From our previous discussion of the thermal compoment of the source function in LTE (Section 5.4) we have

$$S_{\text{therm}}(\tau_\nu, \mu, \nu) = \frac{\varepsilon(\tau_\nu, \mu, \nu)}{\kappa(\tau_\nu, \nu)} = B(\tau_\nu, \nu), \tag{6.23}$$

where B is black body radiation, which depends only on temperature (through τ_ν) and frequency. Because the scattering coefficient σ_{s} is zero, the extincition coefficient is due to absorption only, that is, $\kappa = \alpha$. For thermal emission only in an LTE atmosphere the radiative transfer equation (equation [6.1]) is

$$\mu \frac{dI(\tau_\nu, \mu, \nu)}{d\tau_\nu} = -\kappa(\tau_\nu, \nu) I(\tau_\nu, \mu, \nu) + B(\tau_\nu, \nu). \tag{6.24}$$

We have seen the formal solution to the radiative transfer equation already, equation [5.44]. Using $S = B$, equation [5.44] becomes

$$I(0, \mu, \nu) = \int_0^\infty \frac{1}{\mu} B(\tau_\nu, \nu) \mathrm{e}^{-\tau_\nu/\mu} d\tau_\nu, \tag{6.25}$$

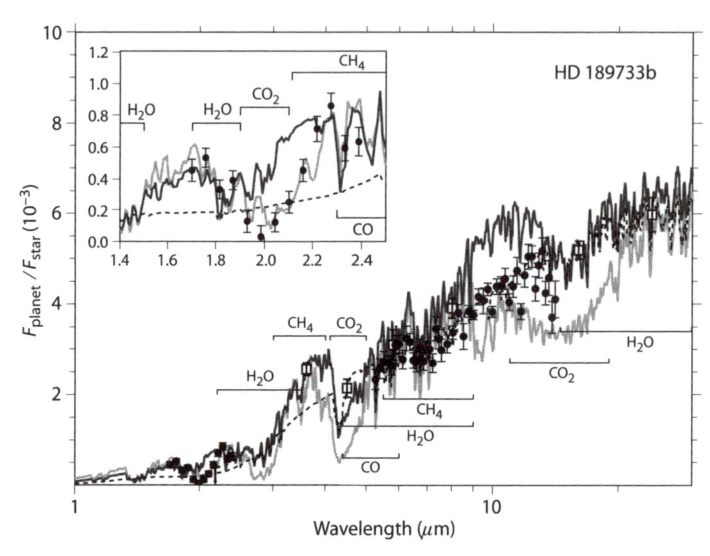

Figure 6.8 Secondary eclipse data for the planet HD 189733b. Three different data sets are shown along with best-fit models. The open squares with error bars show the data from *Spitzer* photometry in six channels [4]. The black circles with error bars in the main panel show *Spitzer*/IRS data from [5]. The black circles in the inset (and the wavelength-corresponding circles in the main panel) show the *HST* NICMOS data from [6]. The curves are model fits (from [2]) to each different dataset.

and the emergent flux is

$$\boxed{F(0,\nu) = 2\pi \int_0^1 \int_0^\infty B(\tau_\nu,\nu)\mathrm{e}^{-\tau_\nu/\mu} d\tau_\nu d\mu.} \tag{6.26}$$

Conceptually, the above equation tells us that the emergent flux is an exponentially weighted black body function where the weighting is the optical depth at a given altitude. Colloquially, we can think of the thermal radiation spectrum as a black body radiation with "bites" taken out of it at frequencies where molecular absorption is occurring. Equation [6.26] is the equation you can use to compute the emergent flux from an exoplanet, and it is powerfully simple because you need only consider a black body in each atmosphere layer and the (more complicated) opacities.

The list of exoplanets with known thermal emission spectra, photometry, or spectro-photometry is growing. In particular, a few dozen planets have been observed by the *Spitzer Space Telescope*, measuring brightness temperatures, identifying presence of molecules, and constraints on exoplanetary vertical temperature structure. Figure 6.8 shows show data for one example, the *Spitzer* and *Hubble Space Telescope* observations of HD 189733b. The thermal emission signal of HD 189733b is up to about 5×10^{-3}. In Chapter 3 we showed that an order-of-magnitude estimate of a planet's thermal emission is $T_{\mathrm{b,p}}R_{\mathrm{p}}^2/T_{\mathrm{b},*}R_*^2$, where the subscript b refers to brightness temperature. The planet-star contrast shown in the

HD 189733 thermal emission spectrum of Figure 6.8 can be used to determine a brightness temperature estimate of the planet.

In the next subsections we make additional simplifying assumptions about the source function to gain further insight into spectral line formation in planetary atmospheres.

6.4.2.2 Constant Source Function

We can solve the radiative transfer equation under the assumption of a constant source function with altitude. See Figure 6.9 for an illustration of this discussion. This subsection follows the useful discussion in [7]. We make the approximation that $S(\tau_\nu, \mu, \nu)$ is constant with optical depth, and for clarity we also drop the μ dependence. Under this approximation the solution to the radiative transfer equation [5.43] becomes

$$I(\tau_{\nu,\rm min}, \mu, \nu) = I(\tau_{\nu,\rm max}, \mu, \nu)\mathrm{e}^{-(\tau_{\nu,\rm max}-\tau_{\nu,\rm min})} + S(\nu)(1-\mathrm{e}^{-(\tau_{\nu,\rm max}-\tau_{\nu,\rm min})}). \quad (6.27)$$

Here the subscript "min" refers to an arbitrary altitude of minimum τ_ν and the subscript "max" refers to an arbitrary altitude of maximum optical depth. From this equation we can see that the intensity at $\tau_{\nu,\rm min}$ has two components. The first component (first term on the right-hand side) is the amount of radiation left over from the original intensity after it has passed through an optical depth $\tau_{\nu,\rm max}$ and been exponentially attenuated. The second component of the emergent radiation (second term on the right-hand side) is the contribution of the radiation emitted along the path traveled by the radiation beam.

First we explore the solution to the radiative transfer equation with a constant source function in the optically thick case. This situation would occur deep in the exoplanet atmosphere where both $\tau_{\nu,\rm min}$ and $\tau_{\nu,\rm max}$ are high and $(\tau_{\nu,\rm max} - \tau_{\nu,\rm min})$ is also high. In this optically thick regime, $\mathrm{e}^{-\tau_\nu} \to 0$ and equation [6.27] becomes

$$\boxed{I(\tau_{\nu,\rm min}, \mu, \nu) = S(\nu).} \quad (6.28)$$

The initial intensity has been completely absorbed and the emergent intensity depends only on the local conditions. In other words, the intensity at $\tau_{\nu,\rm min}$ is equal to the source function. The intensity is unchanged as the beam of radiation travels through the optically thick medium. Deep in the atmosphere LTE is valid. In LTE, the source function is equivalent to black body radiation, $S(\tau_\nu, \mu, \nu) = B(\tau_\nu, \nu)$. At this deep altitude therefore, the intensity is black body radiation—a continuous spectrum with no spectral features. This discussion shows that planetary spectral features do not form at high optical depths.

In contrast to the optically thick regime just discussed, planetary spectra originate in the optically thin regime. We will therefore investigate the case where $\tau_{\nu,\rm max} < 1$ and $\tau_{\nu,\rm min} = 0$ (i.e., at the top of the atmosphere). Consequently $\mathrm{e}^{-\tau_{\nu,\rm max}} \approx 1 - \tau_{\nu,\rm max}$. In this case, the solution to the radiative transfer equation with constant source function equation [6.27] becomes

$$I(0, \mu, \nu) = I(\tau_{\nu,\rm max}, \mu, \nu) - \tau_{\nu,\rm max}\left[I(\tau_{\nu,\rm max}, \mu, \nu) - S(\nu)\right]. \quad (6.29)$$

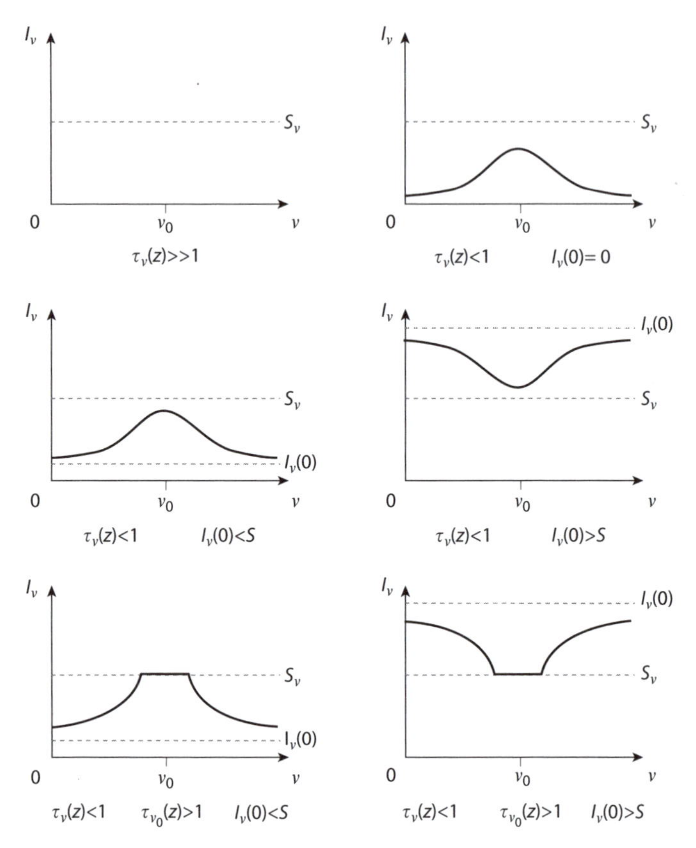

Figure 6.9 Formation of absorption and emission lines for a constant source function with depth and with frequency. Top left: no lines form when the medium is optically thick. Top right: emission lines form when the medium is optically thin and not backlit. Middle left: emission lines also form when the medium is optically thin and $I < S$. Middle right: absorption lines form only when the medium is optically thin and $I > S$. Bottom left and right: the emergent lines saturate to $I(\tau_\nu, \mu, \nu) = S(\tau_\nu, \mu, \nu)$ when the medium is optically thick in the line core but not in the line wings. Following [8].

The emergent intensity $I(0, \mu, \nu)$ depends on the intensity from the bottom of the atmosphere $I(\tau_{\nu,\max}, \mu, \nu)$ diminished or enhanced according to the factor $\tau_{\nu,\max}\,[I(\tau_{\nu,\max}, \mu, \nu) - S(\nu)]$.

Absorption line spectrum. If $I(\tau_{\nu,\max}, \mu, \nu) > S(\nu)$, the initial intensity $I(\tau_{\nu,\max}, \mu, \nu, t)$ will be diminished by the factor $\tau_{\nu,\max}[I(\tau_{\nu,\max}, \mu, \nu) - S(\nu)]$. Absorption lines will therefore be present in the emergent spectrum.

Emission line spectrum. If $I(\tau_{\nu,\max}, \mu, \nu) < S(\nu)$, the initial intensity $I(\tau_{\nu,\max}, \mu, \nu)$ will be enhanced by the factor $\tau_{\nu,\max}[I(\tau_{\nu,\max}, \mu, \nu) - S(\nu)]$. Emission lines will therefore be present in the emergent spectrum.

As a function of frequency, the absorption (or emission) lines will have low (or high) intensity where $\tau_{\nu,max}$ is high. Where $\tau_{\nu,max}$ is high, the absorption coefficient is also high, by the definition of optical depth.

6.4.2.3 Linear Source Function with Depth

At what optical depth do spectral lines form? We can investigate this by approximating the source function as a linear function of optical depth,

$$S(\tau_\nu, \nu) = a(\nu) + b(\nu)\tau_\nu. \tag{6.30}$$

This source function is independent of angle μ. Substituting this form for S into the equation for emergent intensity (equation (5.44)), we find

$$I(0, \mu, \nu) = a(\nu) + b(\nu)\mu. \tag{6.31}$$

With this emergent intensity, the emergent flux can be computed from equation [5.46]:

$$F(0, \nu) = a(\nu) + \frac{2}{3}b(\nu) \equiv S(\tau_\nu = 2/3, \nu). \tag{6.32}$$

In other words, the emergent flux is characteristic of radiation from the $\tau_\nu = 2/3$ region. This is the main point of this subsection. Equation [6.32] is known as the Eddington-Barbier relation. While there are also significant contributions from other optical depths, this linear source function approximation is useful to connect the spectral frequency variation to the atmospheric depth. We emphasize that this result of spectral features generally originating from $\tau_\nu = 2/3$ is not limited to the case of no scattering. As long as a linear source function with altitude holds, this estimate is valid.

If we repeat the previous two steps only for the ray $\mu = 1$, we can estimate the optical depth where the spectral lines form a ray along a $\mu = 1$ path:

$$F(0, \nu) = S(\tau_\nu = 1, \nu). \tag{6.33}$$

In this case, the emergent flux is characteristic of radiation from a $\tau_\nu = 1$ region. We emphasize that τ_ν is frequency dependent. Therefore the physical altitude at which $\tau_\nu = 2/3$ depends on the frequency.

For later use we continue with the zeroth and second intensity moments using the intensity derived from the linear source function with depth, equation [6.31]. We find that

$$\boxed{K(\tau_\nu, \nu) = \frac{1}{3}J(\tau_\nu, \nu).} \tag{6.34}$$

This relationship is known as the Eddington approximation.

6.4.3 A Scattering Atmosphere

So far we have considered solutions to the radiative transfer equation for transmitted radiation and for thermal emission, both cases without scattered radiation. Scattered radiation is responsible for the "reflected" spectrum of a planet at visible (and, for colder planets, at near-IR) wavelengths.

Aside from producing the scattered light spectrum, scattering complicates the solution to the radiative transfer equation both conceptually and computationally. Scattering causes incoming stellar photons to travel long distances before being absorbed to heat the planetary atmosphere. This long-distance photon travel by scattering couples different parts of the atmosphere together, and, for example, causes a global influence of the open boundary at the top of the atmosphere. In contrast to absorption and thermal emission, which couple the radiation field to the local thermodynamic properties of the gas, scattering depends mainly on the radiation field and has ony a weak connection with local values of thermodynamic properties (such as temperature).

The coupling by scattering of one part of the atmosphere to a distant part means that the solution to the radiative transfer equation is complicated. In particular, the radiative transfer equation is no longer a differential equation (equation [6.24]), but becomes an integro-differential equation (see equation [6.40] below). Very few simple analytic solutions are possible. Many textbooks take several pages with long analytic functions to derive expressions for radiation at the planet surface even in the isotropic scattering case. We leave you to read those texts for further information (see the reference section). In this subsection we will describe equations for radiative transfer in an atmosphere that includes scattering, and explore the simplest example that has an analytic solution.

6.4.3.1 Incident Radiation

Let us consider radiation from a star, incident on a planet atmosphere (Figure 6.10). We will take the direction of the star to be θ_0, ϕ_0 away from the surface normal, and the radiation to be plane parallel at the planet surface,

$$I_*(0, \mu, \nu) = I_0 \delta(\mu - \mu_0)\delta(\phi - \phi_0). \tag{6.35}$$

Of later use to us will be the expression for the *unscattered* stellar radiation as it travels down into the planet atmosphere and is attenuated by absorption. At wavelengths where scattering dominates we assume there is no thermal emission and can use the transmitted intensity (equation [6.18]) to find

$$I(\tau, \mu, \nu) = I_0 \delta(\mu - \mu_0)\delta(\phi - \phi_0) \mathrm{e}^{-\tau/\mu}. \tag{6.36}$$

We may also derive the relationship between the incident stellar flux and the incident stellar intensity:

$$\boxed{F_*(0, \mu, \nu) = \int_0^{2\pi} \int_{-1}^{0} \mu I_0 \delta(\mu - \mu_0)\delta(\phi - \phi_0) d\mu d\phi = \mu_0 I_0.} \tag{6.37}$$

6.4.3.2 The Radiative Transfer Equation for Scattering

In a purely scattering atmosphere, scattering comes into play in terms of both extinction and emission. The radiative transfer equation [6.1] is

$$\mu \frac{dI(\tau_\nu, \mu, \nu)}{d\tau_\nu} = I(\tau_\nu, \mu, \nu) + S_{\mathrm{scat}}(\tau_\nu, \mu, \nu). \tag{6.38}$$

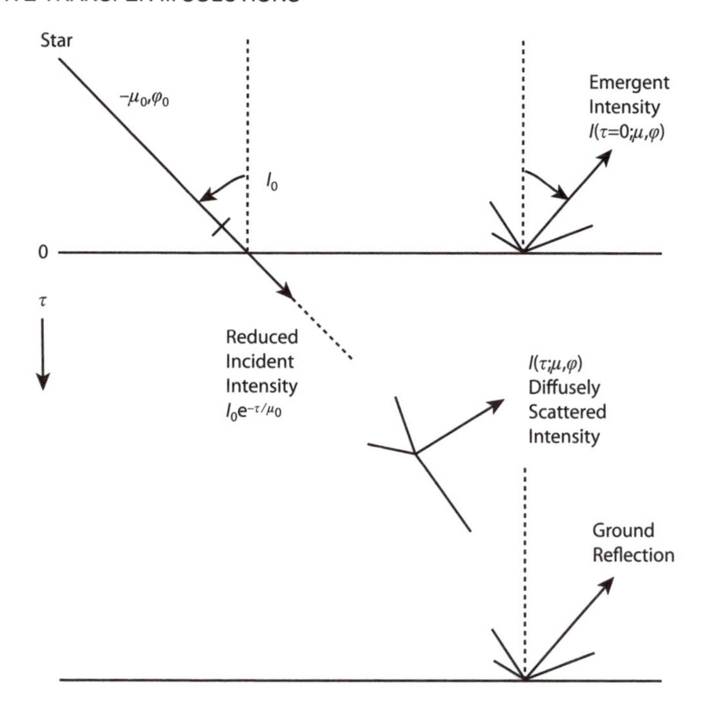

Figure 6.10 Definition of diffuse and scattered radiation. Note the fixed position of the plane parallel incident radiation coming from the star. Adapted from [9].

The scattering source function is $S_{\rm scat} = \varepsilon_{\rm scat}/\sigma_{\rm scat}$, where

$$S_{\rm scat}(\tau_\nu, \mu, \nu) = \frac{1}{4\pi}\int_0^{2\pi}\int_0^{\pi} I(\tau_\nu, \mu, \nu)P(\mu, \phi; -\mu_0, \phi_0)\mu d\mu d\phi, \tag{6.39}$$

so that the radiative transfer equation for a pure scattering atmosphere is an integro-differential equation,

$$\mu\frac{dI(\tau_\nu, \mu, \nu)}{d\tau_\nu} = I(\tau_\nu, \mu, \nu) + \frac{1}{4\pi}\int_0^{2\pi}\int_0^{\pi} I(\tau_\nu, \mu, \nu)P(\mu, \phi; -\mu_0, \phi_0)\mu d\mu d\phi. \tag{6.40}$$

The general form of the emergent intensity is still (from equation [5.44]),

$$I(0, \mu, \nu) = \int_0^{\tau_\nu} S_{\rm scat}(\tau'_\nu, \mu, \nu)\mathrm{e}^{-\tau'_\nu/\mu} d\tau'_\nu, \tag{6.41}$$

which for the scattering atmosphere becomes

$$I(0, \mu, \nu) = \int_0^{\tau_\nu}\frac{1}{4\pi}\int_0^{2\pi}\int_0^{\pi} I(\tau'_\nu, \mu, \nu)P(\mu, \phi; -\mu_0, \phi_0)\mu\mathrm{e}^{-\tau'_\nu/\mu} d\mu d\phi d\tau'_\nu. \tag{6.42}$$

Now we can see why scattering atmospheres are so hard to understand and compute.

In solar system planetary atmosphere studies, the scattering term in the source function is often separated into a single scattering term and a diffuse or multiple scattering term. Diffuse scattering means that incident radiation is scattered in many different directions. The separation of single scattering and diffuse scattering is useful because in some situations the single-scattering term dominates and can be treated alone in a way that is easier to solve. Using the intensity for singly scattered incoming stellar radiation (equation [6.36]), we can write the scattering source function for singly scattered incident stellar radiation (equation [6.39]) as

$$S_{\text{scat,single}}(\tau_\nu, \mu, \nu) = \frac{\tilde{\omega}}{4\pi} I_0 \mathrm{e}^{-\tau/\mu_0} P(\mu, \phi; -\mu_0, \phi_0). \tag{6.43}$$

The source function for diffuse radiation is the same as equation [6.39] above,

$$S_{\text{scat,diff}}(\tau_\nu, \mu, \nu) = \frac{\tilde{\omega}}{4\pi} \int_0^{2\pi} \int_0^{\pi} I(\tau_\nu, \mu, \nu) P(\mu, \phi; -\mu_0, \phi_0, \nu) \mu d\mu d\phi. \tag{6.44}$$

In the last two equations we have used the single scattering albedo $\tilde{\omega}$, relevant in a scattering atmosphere that includes some absorption. Above we had considered a purely scattering atmosphere for which $\tilde{\omega} = 1$.

6.4.3.3 Single Scattering

Let us take the case of a very, very thin atmospheric layer. With a small optical depth, scattering events are dominated by single scattering. If this thin layer is high in the atmosphere, the scattering events are single scattering of the stellar radiation. On Earth this situation occurs in optically thin cirrus and aerosol regions of the atmosphere [10].

The solution to the scattering radiative transfer equation [6.41] with the single scattering source function (equation [6.43]) is

$$I(0, \mu, \nu) = \frac{\tilde{\omega}}{4\pi\mu} \int_0^{\tau_\nu} I_0 P(\mu, \phi; -\mu_0, \phi_0) \mathrm{e}^{-\tau'_\nu \mu_0} \mathrm{e}^{-\tau'_\nu/\mu} d\tau'_\nu, \tag{6.45}$$

leading to

$$I(0, \mu, \nu) = \frac{\tilde{\omega}}{4\pi(\mu + \mu_0)} \mu_0 I_0 P(\mu, \phi; -\mu_0, \phi_0) \left[1 - \mathrm{e}^{-\tau_\nu\left(\frac{1}{\mu} + \frac{1}{\mu_0}\right)}\right]. \tag{6.46}$$

For small τ_ν, we have

$$R(\mu, \phi; \mu_0, \phi_0) = \frac{\pi I(0, \mu, \phi)}{\mu_0 I_0} = \tau \frac{\tilde{\omega}}{4\mu\mu_0} P(\mu, \phi; -\mu_0, \phi_0). \tag{6.47}$$

$R(\mu, \phi; \mu_0, \phi_0)$ is a dimensionless quantity called the bidirectional reflectance function (BRDF). The above equation shows that under the conditions of an optically thin atmosphere, the optical depth is directly proportional to the BRDF. On Earth, the BRDF can be determined from satellite measurements.

We have assumed a very thin layer with single scattering in which the above two solutions for scattered radiation are exact. No assumptions have been made as to the scattering phase function or to the number of angles μ.

6.4.3.4 Multiple Scattering

There are no simple solutions to the radiative transfer equation that includes multiple scattering. One algorithm that is at least straightforward to explain is the

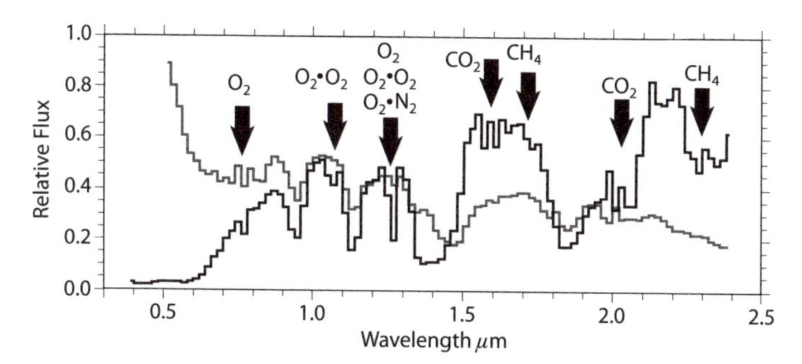

Figure 6.11 Observational comparison of Earth's scattered and transmitted spectrum. The scattered radiation (gray curve) is hemispherically integrated and measured from Earthshine. The transmitted spectrum (black curve) also represents Earth as an exoplanet and was also measured via Earthshine, but using a lunar eclipse. Adapted from [11].

"adding-doubling method." This method uses the reflected and transmitted intensities in many such thin layers added together. Although properties within each layer must be homogeneous, a vertically inhomogeneous atmosphere can be designed by adopting different properties for each layer.

The detection of exoplanets in scattered radiation has proven to be very challenging compared to their detection in thermal emission. One reason is that the group of planets most accesible for atmosphere observations, hot Jupiters, have been found to be very dark, with geometric albedo upper limits on two hot Jupiters at less than 0.15. For comparison, Jupiter's geometric albedo is 0.5 in a comparable wide visible-wavelength bandpass. The directly imaged Fomalhaut planet has an intriguing extended component that is very reflective, but far too large in extent to be attributed to the planetary atmosphere. Finally, scattered radiation studies of Earth as an exoplanet are popular, because we hope to eventually discover and characterize Earth analogs at visible wavelengths We close this subsection with a comparison between Earth's emergent spectrum from scattering and that from transmission, shown in Figure 6.11.

6.4.4 The Diffusion Approximation

We now turn to an approximate solution of the radiative tranfser equation deep in the planetary atmosphere where the conditions of temperature, pressure, and chemical composition change slowly over the length of a mean free path of a photon. This is the case of radiative diffusion and the solution is called the diffusion approximation. Radiative diffusion is the slow, steady movement of photons from one region to another. The diffusion approximation is a description of the radiative flux deep in the atmosphere where $\tau \gg 1$.

We will solve the radiative transfer equation for F under approximations valid deep in the atmosphere. Due to the high density of absorption and scattering particles, the radiation is effectively trapped. In addition, the high particle densities cause collisional rates to dominate over radiative rates, making LTE valid (Section 5.4). Recall that under LTE $S(\tau, \mu, \nu) = B(\tau, \nu)$.

Under the conditions of $\tau \gg 1$ and the source function changing slowly over a distance τ, we can consider a Taylor expansion of $S(\tau', \mu, \nu)$ about a point τ. Approximating the source function by the lowest-order terms, we obtain

$$S(\tau', \mu, \nu) \approx B(\tau, \nu) + \frac{dB(\tau, \nu)}{d\tau}(\tau' - \tau). \tag{6.48}$$

We can solve the radiative transfer equation [6.1] using this source function to find I on $(0 \leq \mu \leq 1)$:

$$I(\tau, \mu, \nu) \approx B(\tau, \nu) - \mu \frac{dB(\tau, \nu)}{d\tau}. \tag{6.49}$$

We emphasize that deep in the atmosphere, as $\tau \to \infty$

$$\boxed{I(\tau, \mu, \nu) = S(\tau, \mu, \nu) = B(\tau, \nu).} \tag{6.50}$$

To summarize, $I = B$ for black body radiation. For thermal radiation, $S = B$. Thermal radiation holds for optically thick media.

We now continue to derive the flux deep in the atmosphere. We follow [12] to put the flux into a diffusion equation form. Recall the definition of flux equation [2.10], and using the above expression for I,

$$F(\tau, \nu) \approx \frac{4\pi}{3}\left[\frac{dB(\tau, \nu)}{d\tau}\right]. \tag{6.51}$$

The flux of any diffusion process is generally described as the product of a diffusion coefficient $\times$ the relevant spatial gradient. For example, heat conduction is written as ($\Phi = -\kappa \nabla T$). We want to find a similar diffusion expression for radiative diffusion—or diffusion flux—in the form of a diffusion coefficient (related to opacity) $\times$ the temperature gradient. We therefore rewrite the above flux equation (equation [6.51]) as

$$F(\tau, \nu) \approx \frac{4\pi}{3}\left[\frac{dB(\tau, \nu)}{d\tau}\right] = -\frac{4\pi}{3}\left[\frac{1}{\kappa_{\mathrm{R}}(\tau, \nu)}\frac{dB(\tau, \nu)}{dT}\right]\left(\frac{dT}{dz}\right). \tag{6.52}$$

This equation is known as the monochromatic diffusion approximation to the radiative transfer equation. Here we have used the Rosseland mean opacity $\kappa_{\mathrm{R}}(\tau, \nu)$ (defined in equation [5.12]).

We can find the total flux by integrating over all frequencies, and using the definition of black body intensity (equation [2.34]),

$$\boxed{F(z) = -\frac{16}{3}\frac{\sigma T^3}{\kappa_{\mathrm{R}}(\tau, \nu)}\frac{dT}{dz}.} \tag{6.53}$$

This expression for total flux is known as the equation of radiative diffusion. The diffusion approximation equation depends only on the temperature, temperature gradient, and Rosseland mean opacity. The term $\frac{16}{3}\frac{\sigma T^3}{\kappa_{\mathrm{R}}}$ is sometimes called the "radiative diffusion coefficient" or the "radiative conductivity" because of the equation's similarity to that for heat conduction presented above. The diffusion approximation equation shows that deep in the planetary atmosphere flux travels in the direction opposite to the temperature gradient.

6.4.5 Radiative Transfer in Cloudy Atmospheres

Up until now we have not mentioned clouds in radiative transfer. In principle, the equations and solutions we have described apply for cloudy atmospheres as well. In practice, numerical solutions are often complicated by the sharp changes in opacity at the boundaries of an optically thick cloud. Most exoplanet atmosphere observations do not yet warrant an accurate treatment of clouds, so a variety of methods to deal with clouds have been used.

Researchers have chosen to deal with cloud boundaries in different ways. Some take the atmosphere surface to be at the layer of an optically thick cloud, entirely avoiding radiative transfer through the cloud for the conceptual reason that photons may scatter indefinitely in a cloud until they are absorbed. Because planets may not be 100% cloud covered, other researchers go a step further and take a weighted average of the intensity emerging from a cloudy and a clear line of sight, for clouds at different altitudes.

Alternatively, for numerical convenience, one may describe the cloud boundary not as a discontinuity but as an exponential density drop-off. Still another approach is to treat the radiation emerging from a cloud layer not by solving radiative transfer through the cloud but by taking an analytic approximation of the scattering of radiation in the cloud before it emerges from the cloud. See [9] for more information.

Ultimately, in a cloudy atmosphere with nonuniform and realistic 3D cloud structures, the Monte Carlo radiative transfer approach is the most accurate way to deal with scattering from clouds.

6.5 MONTE CARLO RADIATIVE TRANSFER

The Monte Carlo radiative transfer technique is a computational technique to solve the radiative transfer equation. We will refer to the previous radiative transfer description as the "traditional radiative transfer technique." There is a fundamental difference between the traditional radiative transfer method for scattered radiation previously described in this chapter and the Monte Carlo method for describing radiative transfer. The former describes a beam of radiation that photons scatter into and out of; the intensity is an aggregate property of photon properties over the whole atmosphere. The Monte Carlo method, in contrast, computationally follows photons as they scatter through the planetary atmosphere. Many, many (millions) of photons need to be followed in order to beat down the noise inherent in use of parameters selected randomly from probability distributions. The main benefit of Monte Carlo models is their ability to deal with inhomogeneous media (e.g., clumpy clouds) and complex geometries. The main challenge is the need to use very large numbers of photons (10^7 or more) in order to sample the probability distribution function space fully, and the computational slowness for optically thick atmospheres. Here we will describe the Monte Carlo technique as applied to scattered radiation only.

The basic principle of the Monte Carlo scattering method is that photon paths and interactions are simulated by random sampling from the various probability distribution functions that determine the interaction lengths, scattering angles, and absorption rates. Incoming photons at a given frequency travel into the atmosphere (to a location sampled from a probability distribution function) and random numbers are used to sample from probabilistic interaction laws for scattering. At each scatter, the photons' polarization and direction change according to the phase function. Photons are followed until they are absorbed (they can no longer contribute to the reflected light), or until they exit the sphere. On exit, the photons are binned into direction and location; the result is the flux and polarization as a function of phase and inclination.

The Monte Carlo description of radiative transfer involves following photons computationally as they travel through a planetary atmosphere. The basic principle of the Monte Carlo method is that the photon paths and interactions are simulated by random sampling from the various physical probability distribution functions that determine the interaction lengths, scattering angles, and absorption rates. The Monte Carlo method is purely computational; there are no equations and hence no analytical solutions. Although this book does not focus on computational methods or solutions, we make an exception to discuss the the Monte Carlo method because it is an amazingly intuitive way to think about how radiation is transmitted through a planetary atmosphere.

The goal of Monte Carlo computations is to use random numbers to sample functions that describe—in a probabilistic manner—physical events. To do this we can use the fundamental principle of Monte Carlo [13] the variable x_0 can be determined uniquely from

$$\zeta = \int_a^{x_0} p(x)dx, \tag{6.54}$$

where ζ is a random number between 0 and 1, and $p(x)dx$ is the probability that x lies between x and $x + dx$. Furthermore, $p(x)$ is normalized in an interval $[a, b]$ such that $\int_a^b p(x)dx = 1$.

A useful example to consider is the distance a photon can travel before an encounter with matter. The probability that a photon will be scattered or absorbed between τ and $\tau + d\tau$ is

$$p(\tau)d\tau = \mathrm{e}^{-\tau}d\tau. \tag{6.55}$$

Substituting this into equation [6.54] and integrating gives

$$\tau = -\log(1 - \zeta). \tag{6.56}$$

To compute a physical distance traveled from the optical depth $\tau(z, \nu)$ (equation [5.14]), we must know the extinction coefficient. In this example of a scattering atmosphere we assume this is known as a function of depth. For the description of radiative transfer, probability distributions other than $\tau(z, \nu)$ must be described as well. These include the probability of absorption or scattering and the directional dependence of scattering. Some of these probability distributions may not have an analytical form of the type $p(x)dx$. In this case, there are many other techniques to sample randomly from distributions.

Because there are no sets of equations it is best to explain Monte Carlo radiative transfer by describing an algorithm for tracing the photon packets as they travel through the planetary atmosphere [13, 14, 15]. For specificity we are assuming a monochromatic, coherent scattering atmosphere, and for comparison with the previous radiative transfer discussions we will focus on a time-independent, homogeneous, plane-parallel atmosphere.

Initialization. We imagine a packet of photons emitted from a source and incident on the planet atmosphere. The photons' location at the top of the planet atmosphere and their direction are probabilistic, depending on the position and distance of the source origin (i.e., the star).

Tracking of photons and their parameters. Once the photons enter the planetary atmosphere they are subject to encounters with gas or solid particles. The distance traveled needs to be computed. The type of encounter, absorption or scattering must be considered. If the photon is absorbed it is assumed to be "killed" because the energy is deposited and reemitted at thermal wavelengths. In other words, the photons are gone from the monochromatic scattering pool. If the photon is scattered, the direction of scattering must be computed. The direction depends on the types of particles available for scattering. This tracking is continued until the photon exits the atmosphere.

Computation of the emergent flux from exited photons. When the photons exit the atmosphere they can be binned into direction. The observable quantity, flux, can be computed from these binned photons.

Errors. Because of the finite number of photons propagated through the planet atmosphere, the number of photons emerging from the planet in each angular bin will obey Poisson statistics. The flux values can therefore be reported with their appropriate error.

The Monte Carlo radiative transfer technique is a rich computational field with applications beyond that of the pure monochromatic scattering and homogeneous atmosphere discussed here. The methods extend to thermal emission models and even computation of temperature-pressure structure.

6.6 SUMMARY

We have explored solutions to the radiative transfer equation. Our goal was to gain a conceptual and quantitative understanding of where planetary spectra come from. We started with a conceptual description of the origin of spectral features, including absorption and emission features in planetary spectra, and their connection to temperature gradients. We also discussed transmission and reflection spectra qualitatively. We investigated quantitative solutions to specific, simplified cases of the radiative transfer equation. These simplifications are sometimes borne out in reality, as in the absence of scattering in a cloud-free atmosphere at red and infrared wavelengths, the absence of thermal emission at visible wavelengths, and the case of transmission valid for transmision spectra where scattering and refraction are usually negligible.

The emphasis of this book is on physical processes, so we did not include a technical description of computation methods. For a fresh alternative to the traditional radiative transfer equation, we described the Monte Carlo solution to radiative transfer, partly because the Monte Carlo method has no equations or analytical solutions. Mostly we described the Monte Carlo method because it is an amazingly intuitive way to think about how radiation is transmitted through the atmosphere.

REFERENCES

For further reading

Radiative transfer in atmospheres is a mature subject and there are many excellent textbooks that discuss spectral line formation and solutions to the radiative transfer equation. For an introductory treatment of many concepts and development of approximate solutions:

- Bohm-Vitense, E. 1989. *Stellar Atmospheres, vol. 2* (Cambridge: Cambridge University Press).

For a thorough study of radiative transfer in a classic treatment of stellar atmospheres:

- Mihalas D. 1978. *Stellar Atmospheres* (2nd ed.; San Francisco: W. H. Freeman).

For a detailed treatment of the scattering atmosphere:

- Sobolev, G. G. 1975. *Light Scattering in Planetary Atmospheres* (New York: Pergamon Press).

For applications to solar system planets:

- Chamberlain, J. W., and Hunten, D. M. 1987. *Theory of Planetary Atmospheres* (New York: Academic Press).

For a thorough treatment of radiative transfer fundamentals and applications to Earth's atmosphere:

- Liou, K. N. 2002. *An Introduction to Atmospheric Radiation* (London: Academic Press).
- Goody, R. M., and Yung, Y. L. 1989. *Atmospheric Radiation: Theoretical Basis* (2nd ed; Oxford: Oxford University Press).

For very detailed and long analytical solutions to various radiative transfer problems:

- Chandrasekhar, S. 1960. *Radiative Transfer* (New York: Dover).

For a detailed description of Monte Carlo computational techniques:

- Cashwell, E. D., and Everett, C. J. 1959. *A Practical Manual on the Monte Carlo Method for Random Walk Problems* (New York: Pergamon Press).

References for this chapter

1. Knutson, H. A., Charbonneau, D., Burrows, A., O'Donovan, F. T., and Mandushev, G. 2009. "Detection of a Temperature Inversion in the Broadband Infrared Emission Spectrum of TrES-4." Astrophys. J. 691, 866–874.

2. Madhusudhan, N., and Seager, S. 2009. "A Temperature and Abundance Retrieval Method for Exoplanet Atmospheres." Astrophys. J. 707, 24–39.

3. Swain, M. R., Vasisht, G., and Tinetti, G. 2008. "The Presence of Methane in the Atmosphere of an Extrasolar Planet." Nature 452, 329–331.

4. Charbonneau, D., Knutson, H. A., Barman, T., Allen, L. E., Mayor, M., Megeath, S. T., Queloz, D., and Udry, S. 2008. "The Broadband Infrared Emission Spectrum of the Exoplanet HD 189733b." Astrophys. J. 686, 1341–1348.

5. Grillmair, C. J., et al. 2008. "Strong Water Absorption in the Dayside Emission Spectrum of the Planet HD 189733b." Nature 456, 767–769.

6. Swain, M. R., Vasisht, G., Tinetti, G., Bouwman, J., Chen, P., Yung, Y., Deming, D., and Deroo, P. 2009. "Molecular Signatures in the Near-Infrared Dayside Spectrum of HD 189733b." Astrophys. J. 690, L114–117.

7. Bohm-Vitense, E. 1989. *Stellar Atmospheres, vol. 2* (Cambridge: Cambridge University Press).

8. R. Rutten Lecture notes, http://www.astro.uio.no/ matsc/school-07/rutten/

9. Chamberlain, J. W., and Hunten D. 1978. *Theory of Planetary Atmospheres* (New York: Academic Press).

10. Thomas, G. E., and Stammes, K., 1999. *Radiative Transfer in the Atmosphere and Ocean* (Cambridge: Cambridge University Press).

11. Palle, E., Osorio, M. R. Z., Barrena, R., Montanes-Rodriguez, P., and Martin, E. L. 2009. "Earth's Transmission Spectrum from Lunar Eclipse Observations." Nature 459, 814–816.

12. Mihalas D. 1978. *Stellar Atmospheres* (2nd ed.; San Francisco: W. H. Freeman).

13. Cashwell, E. D., and Everett, C. J. 1959. *A Practical Manual on the Monte Carlo Method for Random Walk Problems* (New York: Pergamon Press).

14. Code, A. D., and Whitney, B. 1995. "Polarization from Scattering in Blobs." Astrophys. J. 441, 400–407.

15. K. Wood Monte Carlo notes,
http://www-star.st-and.ac.uk/ kw25/research/montecarlo/montecarlo.html

16. Liou, K. N. 2002. *An Introduction to Atmospheric Radiation* (London: Academic Press).

EXERCISES

1. Temperature gradient and spectral features. How does the temperature gradient affect the depth of absorption lines for an LTE atmosphere? In other words, would absorption lines be deeper or shallower, narrower or broader, for a strong temperature gradient compared to a weaker one?

2. Spectral line formation. We introduced the idea that different parts of an absorption line form at different altitudes in the planetary atmosphere (in Section 6.3.2). Sketch an illustration of how different parts of an atomic spectral line form at different altitudes.

3. Transmission spectra.

   a. Describe spectral line formation conceptually for a transmission spectrum.

   b. Derive equation [6.21]. Write down a more accurate expression for the last term on the right-hand side of equation [6.19].

   c. In deriving the transmission spectrum we have assumed that thermal emission in the planet atmosphere makes no contribution. In what case, if any, does thermal emission become relevant?

4. Compare the pathlength through a planet atmosphere for transmitted radiation and thermal emission. Write down a mathematical expression.

5. We have only discussed plane-parallel radiation. One can use plane-parallel results to describe one case on a sphere. Show that the following are equivalent: uniform radiation from one direction onto a unit sphere, and uniform radiation from 2π steradians onto a plane with unit area.

6. A representative value of emergent flux is often taken as $0.5\times$ the outgoing intensity in the direction of the observer, assuming that I is constant. Why?

7. Figure 6.12 shows four components of the radiative transfer equation (after [16]). In this chapter we described the components separately and not all together. The first is stellar radiation attenuated by the atmosphere. The second is singly scattered stellar radiation attenuated by the atmosphere. The third is diffusely scattered radiation. The fourth is thermal emission from an atmosphere layer (from absorbed and thermally processed stellar radiation). Write down the full radiative transfer equation for a small volume of an atmosphere, with each of the four terms. Describe atmospheric conditions where each one would dominate.

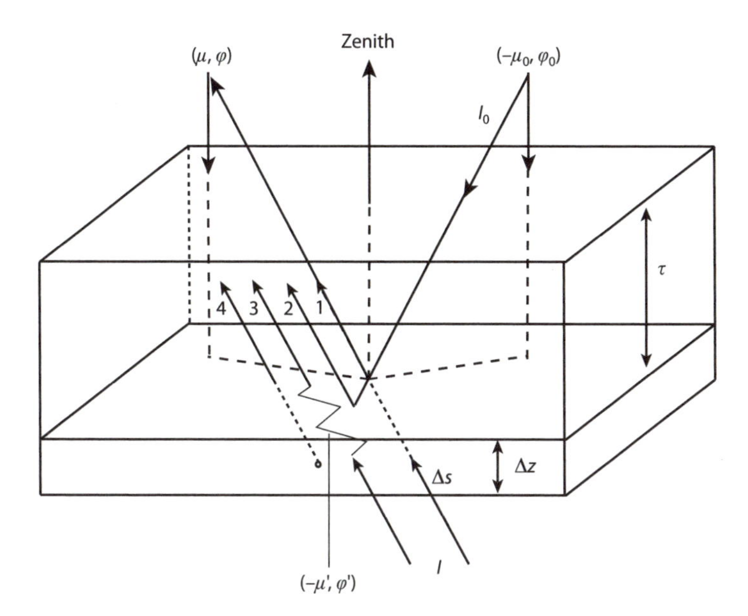

Figure 6.12 Schematic of four atmospheric processes contributing to radiative transfer. (1) Stellar radiation attenuate by the atmosphere. (2) Single-scattered stellar radiation. (3) Diffusely scattered radiation. (4) Thermal emission. See exercise 5. Adapted from [16].

Chapter Seven

Polarization

7.1 INTRODUCTION

We have discussed radiation at length in the previous chapters: we gave a framework to quantitatively describe radiation, a picture of how radiation changes as it travels through a planetary atmosphere, and the concurrent emergent spectrum. Until now, we have omitted all discussion of the "hidden" parameter of radiation, polarization. All electromagnetic radiation can be polarized but is not always.

Polarization is a phenomenon related to transverse waves. A transverse wave is a wave that consists of oscillations (i.e., vibrations) occurring perpendicular (i.e., transverse) to the direction of propagation. While light waves are a collection of waves oscillating in all directions that are perpendicular to the direction of propagation, in polarized light the oscillations occur only in one plane (Figure 7.1).

Examples of polarized light in everyday life abound, although most are hidden from our experience. Sunlight itself is unpolarized, but many kinds of particles can polarize sunlight. Polarized sunglasses are used to block out polarized light, or glare, by not letting horizontally polarized waves through the lenses. Sunlight can be polarized by Rayleigh scattering in Earth's atmosphere. On a clear day you can sometimes see a polarized smudge of light about 90° on either side of the Sun. Polarized sunglasses can be rotated to see the contrast with the adjacent sky increase and decrease. Rainbows are polarized. If you have a camera with a polarizing filter, or polarized sunglasses, you can rotate the polarizer and watch a rainbow disappear and reappear. Solid surfaces, particularly smooth or flat surfaces (such as water or asphalt), can also polarize sunlight.

Radiation is polarized in planetary atmospheres by scattering. In turn, polarization measurements can tell us something about the scattering particles (gas, cloud, or even a surface), beyond information provided by the intensity I alone. As an example, there is a large difference among the polarization functions for gas molecules, cloud, and haze particles. This difference may enable us to tell if a planet is cloud-free or has clouds. For example, if a Rayleigh scattering polarization signature is observed at a range of wavelengths, its cause is likely from atmospheric gas particles. If the planet has clouds, polarization measurements in principle would enable constraints to be placed on the cloud particle size range and composition.

Polarization must be observed as a function of planet orbital phase because the fractional planetary polarization changes with the phase angle (the star-planet-observer angle). It is the features in the polarization "phase curve" that will enable us to identify the composition range and the size range of the scattering particle.

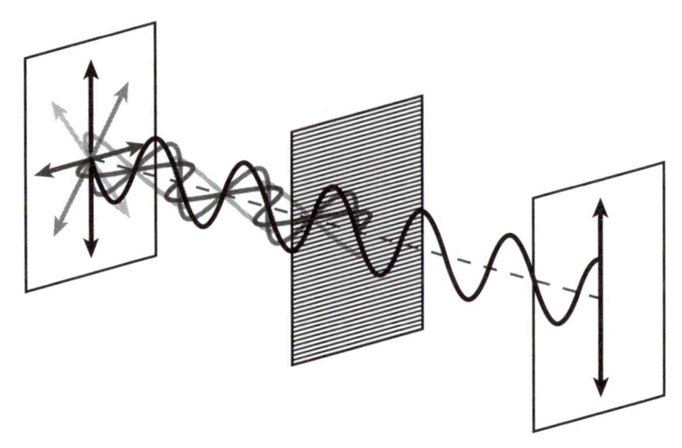

Figure 7.1 Schematic illustration of unpolarized light converted to a linearly polarized beam of radiation by a polarization filter. The arrows show the electric field vector.

Planet atmospheres are linearly polarized and starlight is largely unpolarized. Thus the promise for exoplanet detection and characterization in polarized light: the planet-star contrast ratio is more favorable in polarized than in unpolarized light. As a practicality, however, the polarization signal can be weak. A high intrinsic polarization signature comes from photons that have scattered only once (single scattering). Single scattering, however, means a planet is very dark, because not much light overall is scattered back to space. In contrast, a bright planet is easier to detect, but a bright planet almost always means the photons have undergone multiple scattering. Multiple scattering, unfortunately, causes the polarization to be weak, because for a given phase angle light has been scattered in many different angles, which "washes out" the polarization signal.

7.2 DESCRIPTION OF POLARIZED RADIATION

7.2.1 Basic Concepts

a quantitative description of polarized radiation. To describe polarization, additional parameters other than $I(\tau, \mu, \nu, t)$ are needed. We consider light as a transverse electromagnetic wave with an oscillating electric field. We need to revisit the wave representation of radiation. Electromagnetic radiation is described by a vector wave for the electric field,

$$E(x,t) = A\mathrm{e}^{i(kx-\omega t)}, \tag{7.1}$$

where A is the amplitude, a constant vector for the electric field, x is position, $k = 2\pi/\lambda$ is the wavenumber, ω is angular frequency, and t is time. This electromagnetic plane wave is a solution to Maxwell's equations. Although electromagnetic waves consist of both an electric and a magnetic component, the electric component is responsible for polarized radiation and hence the magnetic compo-

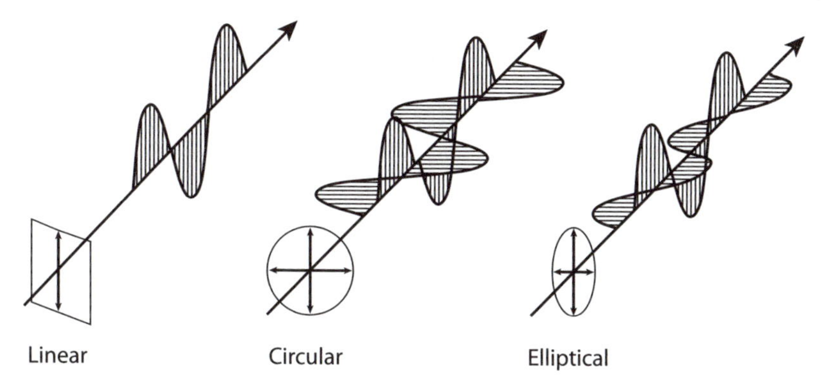

Figure 7.2 Schematic illustration of linear, circular, and elliptical polarization. When viewed looking toward the source, a right circularly polarized beam has a light vector that describes a clockwise circle, while a left circularly polarized beam describes a counterclockwise circle.

nent is typically ignored. Recall that

$$I = \langle E * E^* \rangle, \tag{7.2}$$

where E^* is the complex conjugate of E and the brackets indicate a time average. Light consists of many simple waves where the electric vector has no preferred direction. Polarized light is radiation with the electric field vector oscillating in a fixed direction. Electromagnetic radiation may be partially polarized, and in this case can be described by an unpolarized part and a completely polarized part. The "degree" of polarization refers to the fractional polarization.

Regardless of whether light is polarized or unpolarized, electromagnetic radiation can be broken down into somewhat arbitrary components that are parallel and perpendicular to the plane of wave propagation,

$$E_\perp = A_\perp \mathrm{e}^{i(\omega t - kz - \varepsilon_\perp)}, \tag{7.3}$$

$$E_\parallel = A_\parallel \mathrm{e}^{i(\omega t - kz - \varepsilon_\parallel)}. \tag{7.4}$$

Here, ε describes the phase of the waves.

In this framework, polarized radiation is divided into classes according to the orientation of the electric vector as a function of time (Figure 7.2). Radiation is linearly polarized if the parallel and perpendicular components are in phase (the components may have different relative amplitudes). The tip of the eletric field vector then traces out a single line in the plane. The radiation is circularly polarized if the parallel and perpendicular components have the same amplitude and are exactly 90° out of phase with each other; the electric vector traces out a circle in the plane of propagation. There are two different cases of circular polarization, depending on which direction the field rotates in. These are called right-hand and left-hand polarization. Both linear and circular polarization are special cases of the most general state of polarized radiation: elliptical polarization. Elliptical polarization occurs when the electric field vector traces out an ellipse in the plane of propagation; the two components are not in phase, and are not 90° out of phase,

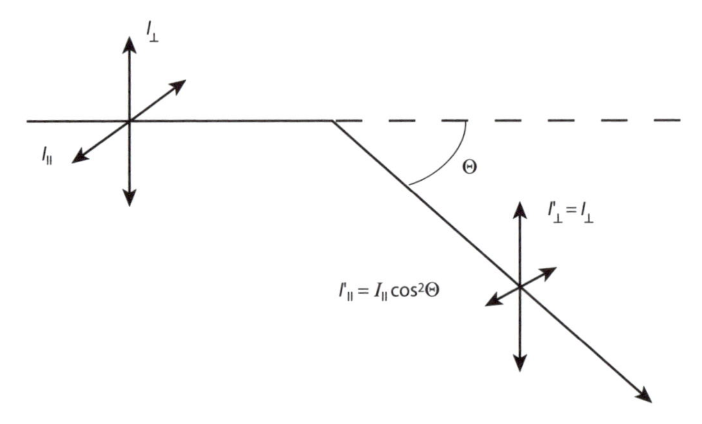

Figure 7.3 Schematic illustration of polarization by Rayleigh scattering. Θ is the scattering angle. The intensity component parallel to the scattering plane is reduced by $\frac{3}{4}\cos^2\Theta$.

and/or do not have the same amplitude. Planetary atmospheres are predominantly linearly polarized.

Four parameters are required for a complete description of polarized radiation. In the parallel and perpendicular description of polarized radiation (equations [7.3] and [7.4]), the four parameters are $A_{\|}$, $A_{\perp}$, $\varepsilon_{\|}$, and $\varepsilon_{\perp}$. These four parameters, are not measurable, however. For observations one might use parameters that are more intuitive: the intensity, the direction of polarization, the degree of linear polarization, and the degree of circular polarization. There are yet other formulations of polarized radiation that we do not describe here. We later describe the Stokes parameters, which are useful for computations and can be directly related to the observables.

this subsection with Rayleigh scattering as an example of an atmospheric scattering process that can polarize radiation. Rayleigh scattering occurs for light scattered by particles that are much smaller than the wavelength of light, such as atomic or molecular gas particles, or even small haze particles. Rayleigh scattering has the polarization $I'_{\perp} = I_{\perp}$, $I'_{\|} = I_{\|}\frac{3}{4}\cos^2\Theta$ where Θ is the scattering angle, $I'_{\perp}$ refers to the intensity perpendicular to the plane of polarization, and $I'_{\|}$ to the intensity parallel to the plane of polarization. Note that $I = I_{\perp} + I_{\|}$, and the degree of linear polarization for single scattering only is

$$P = \frac{I_{\perp} - I_{\|}}{I_{\perp} + I_{\|}} = \frac{1}{2}\left(1 - \frac{3}{4}\cos^2\Theta\right). \tag{7.5}$$

Polarization by Rayleigh scattering is schematically illustrated in Figure 7.3.

7.2.2 The Stokes Parameters

The Stokes parameters are four quantities that describe polarization. Although the Stokes parameters can be related mathematically to the electromagnetic field, they are physically defined by arbitrarily polarized radiation analyzed by four different

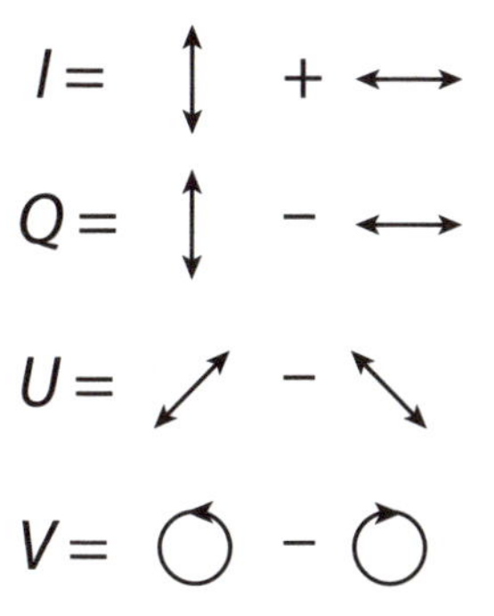

Figure 7.4 Definition of Stokes parameters.

types of polarizer—that is, by sums and differences of the intensity of the radiation field measured along different axes. Conceptually, the Stokes parameters can be represented as shown in Figure 7.4. Our discussion follows an excellent introduction in [1] as well as in several useful textbooks including [2].

Physically, the Stokes parameters represent the following. I, as before, is the total intensity of the radiation field, Q is the difference in intensity between components of the electric vector along two orthogonal directions, U is the difference in intensity between components of the electric vector along two orthogonal axes rotated through 45° from the Q directions, and V is the difference in intensity between the left and right circularly polarized components of the radiation field. Q and U are both necessary in order to prevent vibrations of the electric vector along certain directions from going undetected. For example if the electric vector vibrated only in the (NW-SE) direction, then Q, and hence the polarization, would appear to be zero, when in reality the radiation is 100% polarized. The degree of polarization in terms of the Stokes parameters is

$$P = \frac{\sqrt{Q^2 + U^2 + V^2}}{I}. \tag{7.6}$$

For linear polarization $V = 0$, and for circular polarization $U = Q = 0$.

The Stokes parameters can be mathematically related to and derived from the electric vector description of radiation. Consider a parallel beam of light of circular frequency ω traveling in a certain direction, which we choose to call the positive z direction. The components of the electric field in any two mutually perpendicular directions (represented by unit vectors) may be written in terms of the amplitudes $A_\perp$ and $A_\parallel$ and phases $\varepsilon_\perp$ and $\varepsilon_\parallel$ as described in equations [7.3] and [7.4]. The Stokes parameters are the time averages of the electric vectors

$$I = \langle E_\parallel E_\parallel^* + E_\perp E_\perp^* \rangle = \langle A_\parallel^2 + A_\perp^2 \rangle, \tag{7.7}$$

$$Q = \langle E_\parallel E_\parallel^* - E_\perp E_\perp^* \rangle = \langle A_\parallel^2 - A_\perp^2 \rangle, \tag{7.8}$$

$$U = \langle E_\parallel E_\perp^* + E_\perp E_\parallel^* \rangle = 2\langle A_\parallel A_\perp \cos(\varepsilon_\parallel - \varepsilon_\perp) \rangle, \tag{7.9}$$

$$V = \langle E_\parallel E_\perp^* - E_\perp E_\parallel^* \rangle = 2\langle A_\parallel A_\perp \sin(\varepsilon_\parallel - \varepsilon_\perp) \rangle. \tag{7.10}$$

7.3 POLARIZATION CALCULATIONS

7.3.1 Reflection and Refraction

Polarization can be calculated in a straightforward way with the Fresnel reflection coefficients for radiation in the far field and for particles with large sizes compared to the wavelength of light. The Fresnel equations are for reflection off and refraction through an interface, such as the ocean to air interface. Although the Fresnel equations are not suitable for multiple scattering throughout an atmosphere, they are worth knowing about as a reference [1].

7.3.2 Multiple Scattering

In planetary atmospheres we have described the radiation field using the intensity I, the radiative transfer equation, and the scattering phase function $P(\Theta)$, where Θ is the scattering angle. We now slightly change notation to $P^{11}(\Theta)$, in reference to P as an element of a full phase matrix. In fact, recall the plane-parallel radiative transfer equation [5.33]:

$$\mu \frac{dI(z,\mu,\nu)}{dz} = -\kappa(z,\nu)I(z,\mu,\nu) + \varepsilon(z,\mu,\nu), \tag{7.11}$$

where $\varepsilon = \varepsilon_{\text{scat}} + \varepsilon_{\text{therm}}$, with

$$\varepsilon_{\text{therm}} = \kappa(z,\nu)B(z,\nu) \tag{7.12}$$

and

$$\varepsilon_{\text{scat}}(z,\mu,\nu) = \sigma(z,\nu)\frac{1}{4\pi}\int_0^{\pi}\int_0^{2\pi} P^{11}(\theta,\phi,\theta',\phi')I(z,\theta',\phi',\nu)\sin\theta' d\theta' d\phi', \tag{7.13}$$

where the terms with primes refer to the direction of incidence and the terms with no prime refer to the direction after scattering. Note that here we have written $P(\Theta)$ as $P(\theta,\phi,\theta',\phi')$.

In order to treat polarized radiation in atmospheres with multiple scattering, we must follow four parameters, here the Stokes parameters. The Stokes parameters are intensity-like quantities, and can be treated in the same way as intensity. The Stokes vector is

$$S(\tau,\mu,\nu) = \begin{bmatrix} I(\tau,\mu,\nu) \\ Q(\tau,\mu,\nu) \\ U(\tau,\mu,\nu) \\ V(\tau,\mu,\nu) \end{bmatrix}. \tag{7.14}$$

If I is replaced with the Stokes vector and the formalism for scattered radiation, both the "traditional method" and the Monte Carlo radiative transfer technique otherwise remain the same. To calculate the solution to the radiative transfer equation, we replace I with the Stokes vector and equation [7.11] becomes a set of four coupled differential equations. Many polarization calculations focus on scattered light only, that is, without the term $\varepsilon_{\text{therm}}$.

In order to solve the radiative transfer equation for polarized radiation, we also replace the scalar term $P^{11}(\Theta)$ that describes single scattering with the matrix

$\mathbf{P}(\Theta)$ that describes how the Stokes vector changes under single scattering. Recall that in this traditional radiative transfer description, we treated scattering using the normalized single scattering phase function P^{11},

$$\frac{1}{4\pi}\int_{\Omega} P^{11}(\Theta)d\Omega = 1. \tag{7.15}$$

The phase matrix is similarly normalized.

The full phase matrix can be represented by

$$\mathbf{P} = \begin{bmatrix} P^{11} & P^{12} & P^{13} & P^{14} \\ P^{21} & P^{22} & P^{23} & P^{24} \\ P^{31} & P^{32} & P^{33} & P^{34} \\ P^{41} & P^{42} & P^{43} & P^{44} \end{bmatrix}. \tag{7.16}$$

A simplification is possible for Rayleigh scattering and Mie scattering (and a few other cases; Chapter 8) in that the phase matrix has no more than six independent phase matrix elements

$$\mathbf{P} = \begin{bmatrix} P^{11} & P^{21} & 0 & 0 \\ P^{21} & P^{22} & 0 & 0 \\ 0 & 0 & P^{33} & P^{-43} \\ 0 & 0 & P^{43} & P^{44} \end{bmatrix}. \tag{7.17}$$

For spherical particles, the phase matrix elements depend only upon the particle index of refraction and the particle size compared to the wavelength of light. See Chapter 8 for a further description of the source of the scattering matrix elements. For now, we will emphasize that the single scattering phase function is P^{11} and the fractional polarization for single scattering (sometimes called the linear polarization probability function) is $-P^{21}/P^{11}$.

The phase matrix is defined with respect to the plane of scattering, where Θ is the scattering angle. In multiple scattering problems there are many different scattering planes. To choose a common reference plane for the Stokes parameters, we follow [1, 3] and define the reference plane by the emergent direction (i.e., the direction to the observer) and the local normal ($\mu = \cos\theta = 1$). The phase matrix must be rotated into and out of the reference frame, whereby the Stokes vector after scattering through an angle Θ is

$$S = \mathbf{L}(\pi - i_2)\mathbf{P}\mathbf{L}(-i_1)S'. \tag{7.18}$$

Here $\mathbf{L}$ is a Mueller rotation matrix,

$$L(\psi) = \begin{bmatrix} 0 & 0 & 0 & 0 \\ 0 & \cos\psi & -\sin\psi & 0 \\ 0 & \sin\psi & \cos\psi & 0 \\ 0 & 0 & 0 & 0 \end{bmatrix}, \tag{7.19}$$

and, for definitions of the angles i_1 and i_2 in terms of the scattering angle Θ, see Figure 7.5. The rotation angles i_1 and i_2 are given by

$$\cos i_1 = \frac{\mu(1-\mu'^2)^{1/2} - \mu'(1-\mu^2)^{1/2}\cos(\phi - \phi')}{(1-\cos^2\Theta)^{1/2}}, \tag{7.20}$$

$$\cos i_2 = \frac{\mu'(1-\mu^2)^{1/2} - \mu(1-\mu'^2)^{1/2}\cos(\phi - \phi')}{(1-\cos^2\Theta)^{1/2}}, \tag{7.21}$$

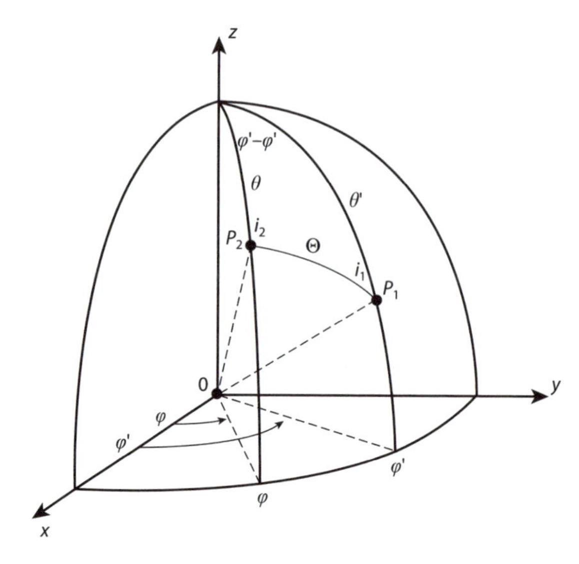

Figure 7.5 Definition of angles used in the Mueller rotation matrix. Radiation traveling into direction $P_1(\theta', \phi')$ scatters through angle Θ into direction $P_2(\theta, \phi)$. Adapted from [4].

and the scattering angle Θ is

$$\cos\Theta = \mu\mu' + (1-\mu^2)^{1/2}(1-\mu'^2)^{1/2}\cos(\phi - \phi'). \tag{7.22}$$

The Stokes polarization matrix elements are derived from Mie theory.

7.4 POLARIZATION FROM PLANETS

Polarization measurements of planets require observations as a function of orbital phase, with as much phase coverage as possible. The reason is that the polarization signature lies in polarization features at specific phases (see ahead to Figure 7.8). It is the polarization changes with phase that enable us to identify the scattering particle type and size. Many people think of Rayleigh scattering, with a peak polarization at 90°. Yet condensate particles have features at other phases (Figure 7.6).

We emphasize a difference in reflected light phase curves and polarization phase curves that has to do with normalization. The planet flux (i.e., brightness) change is mostly due to the changes in the illumination geometry of the planet. As a function of orbital phase, a planet in scattered light gets brighter toward "full phase" and dimmer toward "crescent phase." For polarized light, the polarization phase curve is reported as a ratio of polarized flux to total planet flux, so that such geometric effects cancel out. The polarization phase curve features then depend on on the properties of the scattering particles. See Figure 7.7 for a comparison of flux and polarization peaks for Rayleigh scattering.

We turn to two fundamental points related to polarization of planetary atmospheres. First, for many cases of scattering atmospheres, the full description of

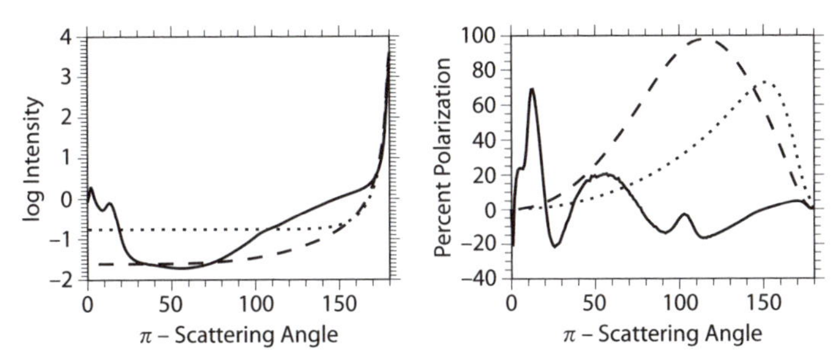

Figure 7.6 Single scattering phase function (left) and linear polarization probability function (right) for three high-temperature condensates: $MgSiO_3$ (solid), Fe (dotted), and Al_2O_3 (dashed). Linear polarization is more sensitive to composition: most of the distinctive features in the scattering phase function (left) are at quite low relative intensities. Adapted from [5].

radiation using the four Stokes parameters is needed to calculate an accurate emergent intensity. Computation of the intensity alone emerging from a scattering atmosphere contain errors on the order of 10% or less for Rayleigh scattering [3] and errors on the order of 1% or less for particles with sizes comparable to or larger than the wavelength of light [1]. A second fundamental point is that for planetary atmospheres linear polarization dominates and for solar system planets, circular polarization has been measured to be very small.

Venus is a canonical example of how a polarization measurement changed our view of a planet atmosphere (Figure 7.8). Venus has 100% cloud cover and is very bright, with a geometric albedo of 0.65. Historically, people assumed Venus's clouds were made of liquid water, just as some of Earth's clouds are. This assumption dates back to the sentiment that Venus and Earth are sister planets and should have much in common. Even with flux measurements as a function of orbital phase, the size range and particle composition still included liquid water droplets, although other species would also have been possible. With polarization measurements taken at a range wavelengths in the late 1960s and earlier, Venus's cloud particles were definitively identified to be hydrosulfuric acid (H_2SO_4) clouds with a mean particle size of approximately 1 μm. This identifcation was later confirmed in situ by the *Pioneer Venus* in the late 1970s and 1980s.

The identification of Venus's cloud particles was a triumph for polarization measurements. Venus' clouds consist of spherical liquid droplets, which makes polarization calculations relatively straightforward. In contrast to liquid droplets, solid particles are not usually spherical. It is very challenging to compute the phase matrix for nonspherical particles, and the range of shapes in a given ensemble of cloud particles means many free parameters. With the difficulty of modeling scattering and polarization of solid particles, it may be more difficult to interpret polarization data on exoplanets.

For exoplanets we can use the percent polarization as a guide to what a polarization signature as a function of phase might look like. If single scattering dominates in a planet atmosphere, the planet's polarization signal will look like that

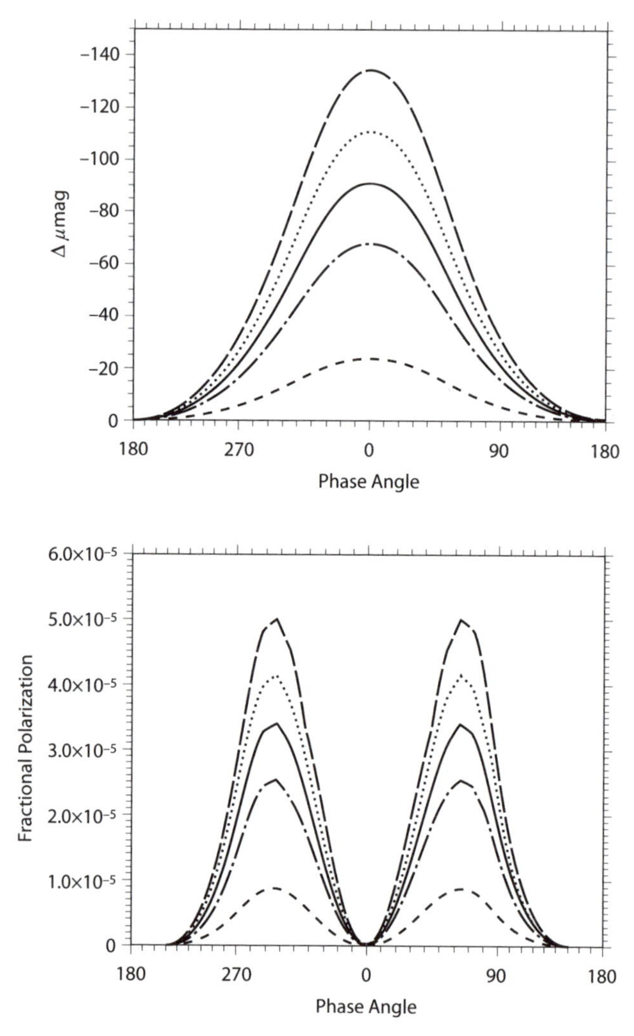

Figure 7.7 Scattered light (top) and fractional polarization (bottom) for Rayleigh scattering. $\Delta\mu$mag is micro magnitudes in the stellar magnitude system.

shown in Figure 7.6. In reality, multiple scattering is more likely to occur in a planet atmosphere and will wash out the signature, sometimes to an extreme. At the present time, astronomers are focused on detecting polarized light signatures from hot Jupiters, by observing the combined light of the planet-star system.

One of the most intriguing ideas for exoplanet polarization measurements is that of polarization from specular reflection off an Earth-like exoplanet's oceans. Specular reflection is a mirrorlike reflection from a surface and depends on the direction between the Sun (or star) and the observer. You can see specular reflection off oceans or lakes in the form of bright glints off waves. Space-based observations of Earth show the specular reflection glint off the oceans, but in the hemispherically

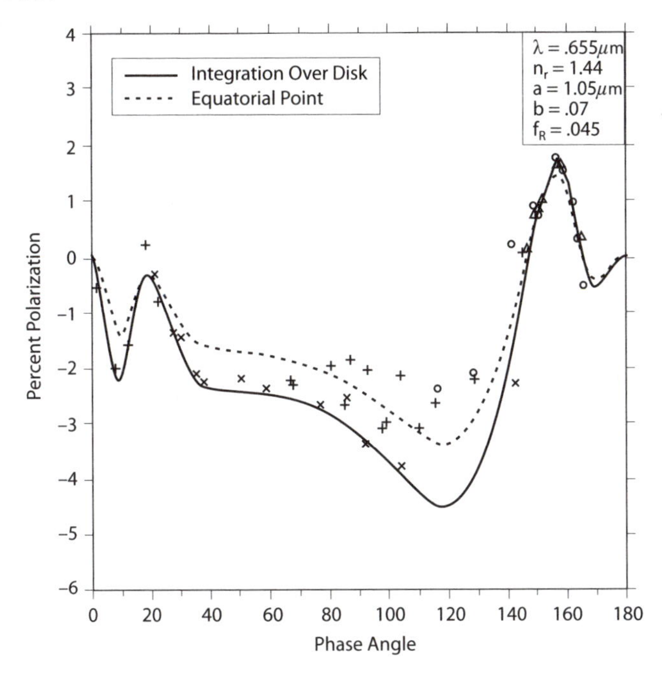

Figure 7.8 Venus's atmospheric polarization. Adapted from [6], and see references therein.

averaged Earth the specular reflection glint does not stand out. The glint should be polarized, and so the promise in detecting the glint on a distant planet is the hope of identifying a smooth surface—one that, based on the equilibrium temperature of the planet can be attributed to liquid water oceans. Contamination from atmospheric polarization may make this measurement difficult. Polarization from ocean specular reflection may nevertheless be a way of identifying liquid water oceans on Earth-like exoplanets in the distant future when the requisite technology exists.

7.5 SUMMARY

Polarization is an interesting topic for exoplanets for two reasons. First, polarization measurements have the potential to yield much more information about scattering particle sizes and composition than can be obtained through scattered light or thermal emission alone. Second, the facts that planet atmospheres are linearly polarized and starlight is unpolarized create a promising planet-star polarization contrast for detection. Nevertheless, polarization signals will be low and challenging to detect.

For solar system planets polarization enabled the remote identification of the Venusian cloud particles, breaking a degeneracy that included liquid water droplets. In situ measurements confirmed that the clouds are made up of small hydrosulfuric acid particles—very different from liqud water. As a practical discussion, we presented the Stokes parameters as a formal way to follow polarized radiation through the radiative transfer equation.

REFERENCES

For further reading

A thorough article that includes polarization fundamentals for atmospheres:

- Hansen, J. E., and Travis, L. D., 1974. "Light Scattering in Planetary Atmospheres." Space Sci. Rev. 16, 527–610.

A fundamental textbook that includes polarization:

- van de Hulst, H. C. 1981. *Light Scattering by Small Particles* (New York: Dover).

References for this chapter

1. Hansen, J. E., and Travis, L. D., 1974. "Light Scattering in Planetary Atmospheres." Space Sci. Rev. 16, 527–610.
2. van de Hulst, H. C. 1981. *Light Scattering by Small Particles* (New York: Dover).
3. Chandrasekhar, S. 1950. *Radiative Transfer* (Oxford: Oxford University Press).
4. Code, A. D., and Whitney, B. 1995. "Polarization from Scattering in Blobs." Astrophys. J. 441, 400–407.
5. Seager, S., Whitney, B., and Sasselov D. D. 2000. "Photometric Light Curves and Polarization of Close-In Extrasolar Giant Planets." Astrophys. J. 540, 504–520.
6. Hansen, J. E., and Hovenier, J. W. 1974. "Interpretation of the Polarization of Venus." J. Atm. Sciences 31, 1137–1160.

Chapter Eight

Opacities

8.1 INTRODUCTION

The way in which radiation moves through a planet atmosphere is controlled by opacities: the numbers of molecules, atoms, or condensates present and their absorbing characteristics. By absorbing and emitting stellar radiation, the gas and solid opacities have deep consequences for so many of the physical characteristics of the atmosphere (see Figure 8.1). The planet atmosphere temperature structure depends on where and how much of the incoming and outgoing radiation is absorbed, scattered, and thermally reradiated. Whether or not convection occurs depends in part on the magnitude of opacities; if radiative energy transport is impeded by the opacities, then energy transport by convection will set in. The planetary spectrum arises from absorption and emission of radiation. Most significantly for physical characterization of exoplanets, the planetary spectrum is one of the most important observables.

In this chapter we address the fundamentals of opacities: molecular energy levels and their populations, and absorption cross sections. We focus on molecules and not atoms. Due to planetary temperatures, molecules dominate observable exoplanetary spectra. We aim to understand why the absorption bands of an individual molecule are located at a given frequency and with a given line strength. We also strive for a basic picture of absorption and scattering from solid particles. Condensates—solid particles congregating as clouds or haze layers—can have strong opacities and are expected to form in many planetary atmospheres.

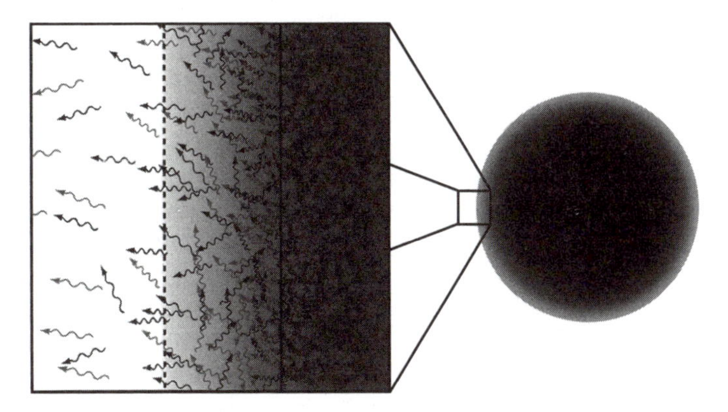

Figure 8.1 Illustration of photons emerging from a giant planet atmosphere. Based on [1].

As an introduction to opacities we review the expression for the absorption coefficient α (equation [5.9] from Section 5.2.4),

$$\boxed{\alpha(T,P,\nu) = \sum_j \sum_i n_{ji}(T,P)\sigma_{ji}(T,P,\nu).} \tag{8.1}$$

The absorption coefficient is made up of contributions from each gas species j. In Chapter 4 we described chemical equilibrium, nonequilibrium, and photochemistry calculations that led to the number densities (in units of m^{-3}) of different molecules. For each kind of molecule j, there are contributions from molecules with different energy states i. More specifically, the term n_{ji} refers to the number density of molecules j in energy state i. The number density of molecules is determined by chemical reactions at the ambient temperature and pressure in a specific atmosphere layer (Section 4.3) and by cloud processes (Section 4.4).

The focus of the first part of this chapter is on the energy levels and absorption cross sections for transitions between energy levels. In the second part, we turn to describe the emission coefficient, with a focus on scattering from condensates.

8.2 ENERGY LEVELS IN ATOMS AND MOLECULES

Atoms and molecules have many individual energy levels that are characteristic to the specific atom and molecule. From quantum mechanics, we know that atoms and molecules have discrete energy states, unlike classical particles which can have any energy. The difference between two energy levels corresponds to the wavelengths of spectral features via the relation $\Delta E = h\nu$, where h is Planck's constant and $h\nu$ is the amount of energy absorbed by or carried away from an atom or molecule by a photon.

For planetary atmospheres we are interested in the emergent spectrum. Planetary scientists use wavelength, frequency, or wavenumber to describe the energy of spectral features, and not the energy of the atomic or molecular transition itself. The wavenumber of electromagnetic radiation is defined as

$$\tilde{\nu} = \frac{1}{\lambda} = \frac{\nu}{c} = \frac{E}{hc} \tag{8.2}$$

in units of m^{-1} or more conventionally cm^{-1}. Wavenumber is more often used among spectroscopists than wavelength because energy levels are conveniently proportional to wavenumber (or frequency) but inconveniently inversely proportional to wavelength. Similarly, spectrometers are often calibrated in wavenumber because it is independent of the fundamental constants c and h. The wavenumber in spectroscopy should not be confused with the angular or circular wavenumber used to described waves in the wave equation. When describing a spectrum, wavelength, frequency, and wavenumber are often used interchangeably. We will either use $\tilde{\nu}$ to denote wavenumber, or make it clear whenever ν refers to wavenumber.

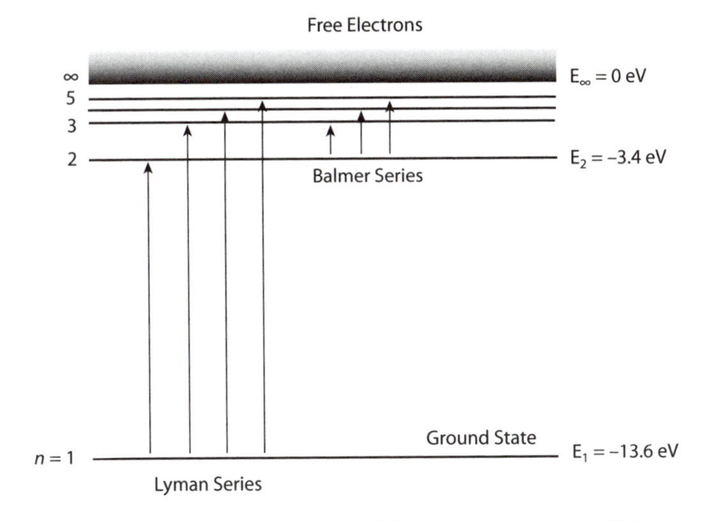

Figure 8.2 Hydrogen energy level diagram. See equation [8.3].

8.2.1 Atoms

Let us start with the Bohr model of the hydrogen atom as a way to recall some of the basics of atomic energy levels. The Bohr model for an isolated atom is that of a postively charged nucleus with electrons orbiting in successive orbits. Bohr developed the basic picture that electrons in an atom have stationary states, or energy levels, and that when an electron changes its energy level a photon is emitted or absorbed. Bohr used the equation of motion of an electron, including the kinetic energy and Coulomb potential energy terms, to derive the total energy state of an atomic system:

$$E_n = -\frac{me^4}{8\epsilon_0^2 h^2}\frac{1}{n^2} = -\frac{R_H hc}{n^2}, \tag{8.3}$$

where $n = 1, 2, \ldots \infty$. Here m is the mass of the electron, e is the electron charge, ϵ_0 is the electric constant, and R_H is Rydberg's constant. n is now known as the principal quantum number. From equation [8.3], we can directly calculate the energy levels of the hydrogen atom (see Figure 8.2) and calculate the wavenumber or frequency of hydrogen energy transitions, that is, the wavenumber of hydrogen line spectral features. Bohr also postulated that the angular momentum L can take on only discrete values:

$$L = n\frac{h}{2\pi}. \tag{8.4}$$

Bohr reached his model and equations from experimental evidence and heuristic arguments, motivated by the desire to explain the hydrogen gas spectrum and frequencies. To proceed to atoms with more than one electron, quantum mechanical equations are needed. We will therefore review a few concepts from quantum mechanics. Although we will not solve any equations in this chapter, our aim is for a conceptual understanding of the pathway to determine the energy levels of atoms and molecules.

Recall that for classical mechanics the Hamiltonian is used as a description of the kinetic and potential energy, that is, the total energy, of a closed system. In quantum mechanics, the Hamiltonian is the operator associated with the kinetic and potential energies of a system—in quantum mechanics each measurable parameter has an associated operator. The Hamiltonian operator is used to determine the energies of a system. Operating on the wavefunction, in the time-independent case, the Hamiltonian operator generates the time-independent Schrodinger equation which can be solved to find the energy eigenvalues (i.e., energy levels) of the system:

$$H\psi_i = E_i\psi_i, \tag{8.5}$$

where ψ are eigenfunctions known as the stationary wavefunctions, the probability amplitudes for the particle to have a given position x at time t. E_i are the energy eigenvalues, and i refers to individual levels. H is the Hamiltonian operator associated with the kinetic and the potential energy and in one dimension is

$$H = \frac{-\hbar^2}{2m}\frac{\partial^2}{\partial x^2} + V(x), \tag{8.6}$$

where x is position. Here we have taken the kinetic energy as $p^2/2m$ where p is momentum and m is mass. We have used $\frac{d^2\psi}{dx^2} = -\frac{p^2}{\hbar}\psi$ and considered wavefunctions of the form $\psi(x,t) = Ae^{i(kx-\omega t)}$. The 1D time-independent (nonrelativistic) Schrodinger equation is then

$$\boxed{\frac{-\hbar^2}{2m}\frac{d^2\psi(x)}{dx^2} + V(x)\psi(x) = E\psi(x).} \tag{8.7}$$

The energy levels of the hydrogen and other atoms can be derived from Schrodinger's equation by specifying $V(x)$, that is, how the potential energy changes with location. For the hydrogen atom, one can treat the potential energy term as a classical Coulomb potential,

$$V(r) = \frac{e^2}{4\pi\epsilon_0}\frac{1}{r}, \tag{8.8}$$

where r is the radial distance of the electron from the force center and in equation [8.7] r replaces x and m is the reduced mass of the system. For hydrogen and hydrogen-like ions, the energy levels can be computed analytically. For many-electron atoms, the potential is more complicated and computers are needed to work out solutions to the Schrodinger equation.

8.2.2 Molecules

Conceptually, a molecule can be visualized as number of atoms bound together by a balance of mutually attractive and repulsive forces. Molecules undergo electronic transitions just as atoms do. In electronic transitions, the electron moves from a lower to a higher energy level upon absorption of a photon, and the molecule emits a photon as the electron drops to a lower energy level. An electronic transition typically involves a few eV ($\sim 10^4$ cm^{-1}) of energy. This transition energy corresponds to photons in the UV or visible range of the electromagnetic spectrum.

Molecules, unlike atoms, have energy transitions other than electronic transitions. These rotational and vibrational transitions lead to complex band systems. The molecule as a whole rotates about any spatial axis, giving rise to rotational transitions. The individual atoms vibrate with respect to one another, causing vibrational transitions. When vibrational motion occurs, rotational motion is always induced. The water vapor molecule, for example, has hundreds of millions of lines from combined rotational-vibrational transitions. The molecular spectral lines resulting from transitions between energy levels blend together to form molecular bands.

For rotational or vibrational radiative energy transitions, the molecule must couple with an electromagnetic field so that energy exchanges can take place. This coupling is generally provided by the electric dipole moment of the molecule. A dipole moment in a molecule exists if the effective centers of the positive and negative charges of the molecule have nonzero separation. Essentially, a molecule has a dipole moment if it has a difference between the center of charge and the center of mass. For example, H_2O and O_3 have permanent electric dipole moments due to their asymmetrical charge distributions. Linear molecules, such as N_2, O_2, and CO_2 are examples of molecules that have no permament dipole moment because of their symmetrical charge distributions,. Lack of a permanent dipole moment theoretically means the gas species has no rotational-vibrational transitions. Dipole moments can be temporarily induced. For example, asymmetric bending or stretching modes of vibration induce a dipole moment (e.g., CO_2). In addition, the weaker electric quadrupole or magnetic dipole moments may exist and cause vibrational transitions.

To get some perspective on energy levels in molecules we show a potential energy diagram for molecular transitions in Figure 8.3. Two stable electronic states of a diatomic molecule are shown. Each electronic state has several vibrational states of lower energy associated with it. Each vibrational state, in turn, has many even lower-energy rotational states associated with it.

To quantitatively describe molecular rotational and vibrational energy levels we will adopt a conventional, pedagogical, yet mechanical model of the molecule. This model ignores the detailed structure of the molecule in terms of nuclei and electrons, but enables us to get a handle on the origin and structure of molecular lines.

8.2.2.1 Molecular Rotational Transitions

We will begin with the pure rotational spectrum of a diatomic molecule. To gain a conceptual understanding of the origins of the rotational energy levels, we take the conventional approximation of a rotating diatomic molecule as a rigid rotor. In this picture, the atoms are the two masses with a fixed separation and rotating about an axis that goes through the common center of mass (Figure 8.4.) We are assuming that the diatomic molecule can rotate about the common center of mass without changing the separation or relative positions of the constituent atoms.

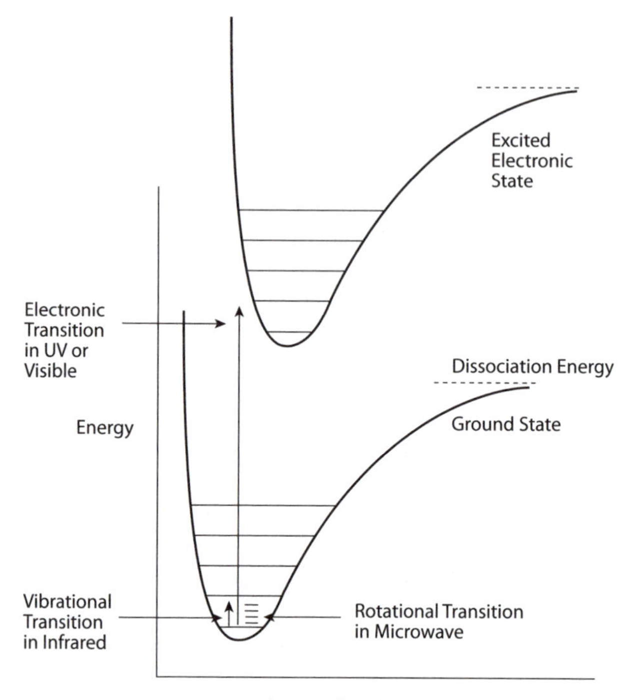

Figure 8.3 Schematic illustration of molecular electronic, vibrational, and rotational energy levels. Note that near the potential energy minimum of an electronic energy state, the potential energy curve can be approximated by a parabola in the same functional form as the simple harmonic oscillator potential. Adapted from [2].

Our aim is to understand where the rotational energy levels of a diatomic molecule come from. We will use the time-independent Schrodinger equation framework presented in Section 8.2.1, and find the eigenvalues for the 1D time-independent Schrodinger equation with the relevant kinetic and potential energy terms. In fact, a rigid rotor has no potential energy. We are literally making the assumption that the atoms in the diatomic molecule are completely fixed with respect to each other.

Classically, a rigid rotor with angular velocity ω has moment of inertia[1] I.

$$I = \sum_i m_i r_i^2 = m_1 r_1^2 + m_2 r_2^2, \tag{8.9}$$

where r_1 and r_2 are the distances of the atoms from the common center of mass. The angular momentum of a classical rigid rotor is

$$L = \sum_i m_i \omega r_i. \tag{8.10}$$

[1] Here the moment of intertia I is not to be confused with the intensity I used to describe radiation.

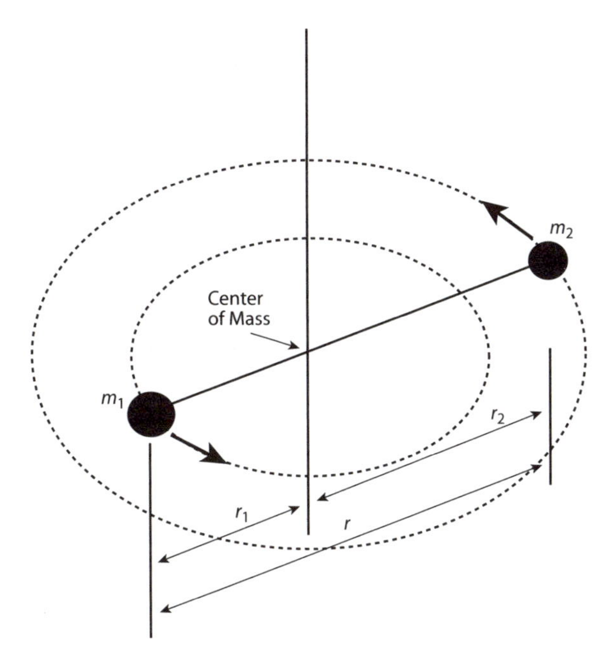

Figure 8.4 Schematic illustration of the rigid rotor. The rigid rotor is used as an approximate model for a diatomic molecule.

The classical rotational kinetic energy from a rigid rotor is therefore

$$E = I\omega^2 = \frac{L^2}{2I}. \tag{8.11}$$

We may now associate the classical rotational energy term with the quantum mechanical Hamiltonian operator,

$$H\psi = \frac{L^2}{2I}\psi = E_J\psi \tag{8.12}$$

or

$$L^2\psi = 2IE_J\psi. \tag{8.13}$$

The wavefunction solutions ψ to this equation are actually the spherical harmonics $Y_{l,m}(\theta, \phi)$. If we were to put the spherical harmonics into equation [8.13], then we would find

$$L^2 Y_{l,m}(\theta,\phi) = J(J+1)\hbar^2 Y_{l,m}(\theta,\phi) = 2IE_J\psi \tag{8.14}$$

or

$$\boxed{E_J = J(J+1)\frac{\hbar^2}{2I}} \tag{8.15}$$

with rotational quantum number $J = 0, 1, 2, \ldots$. Often the rotational energy levels are written as

$$E_J = BJ(J+1), \tag{8.16}$$

with B defined as the rotational constant,

$$B = \frac{\hbar^2}{2I}. \tag{8.17}$$

If we were to solve equation [8.14], we would find the degeneracy g, the number of states with the same energy but different quantum numbers,

$$g_J = 2J + 1. \tag{8.18}$$

The rotational energy levels have spacing that increases with J. Interestingly, the energy transitions are constant in wavenumber or frequency, meaning that rotational lines are equally spaced in a wavenumber or frequency spectrum. We can see this by considering that allowed rotational transitions have $\Delta J = \pm 1$ for linear molecules (see Section 8.3.2). Then,

$$\Delta E = E_J - E_{J-1} = 2BJ. \tag{8.19}$$

We can use equation [8.15] to determine the order-of-magnitude energy of pure rotational spectra of diatomic molecules. Let us use the diatomic molecule CO as an example. To first order, the C and O atoms have similar atomic masses, $m_1 \sim m_2 \approx 12$ amu, and the internuclear separation in diatomic molecules is about $r_1 \sim r_2 \approx 1$ Å. We find that $E_J \sim 1.4 \times 10^{-23}$ J. This corresponds to energies at microwave frequencies: a frequency of 2×10^{10} s^{-1}, a wavenumber of 0.7 cm^{-1}, or a wavelength of 0.014 m. A pure rotational spectrum can exist at these very low energies—energies too low to excite vibrational modes of the molecule.

A complication to the rigid rotor approximation for linear diatomic molecules is that molecules are not completely rigid. The centrifugal force means that the atomic nuclei will respond to increasing rotational energy by moving further apart. The resulting increase in internuclear separation with increasing rotational quantum number J decreases the rotational energy level separation. To account for this centrifugal elongation, an extra term must be added to the rotational energy level equation. The centrifugal distortion is significant typically only for very highly rotationally excited states.

We developed the rigid rotor framework for linear diatomic molecules. The discussion is also valid for spherical tops (or any rigidly rotating dipole). For molecules that are asymmetric tops, an additional term is required.

In this section, we have not solved the Schrodinger equation, but rather asserted a solution. To gain a deeper understanding, one could follow a general approach taken in mathematical physics by formulating a guess at a solution based on solutions of similar equations. For a much more detailed derivation of molecular energy levels, there are many textbooks [e.g., 3, 4] and reference books [e.g., 5] to name a few. We will now continue on to molecular vibrational energy levels, using the same classical analogy approach we took for the rotational energy levels.

8.2.2.2 Molecular Vibrational Transitions

Molecules vibrate when the atoms in a molecule are in periodic motion with respect to the molecule's center of mass (Figure 8.5). Molecular vibrational transitions never occur on their own; as the molecule vibrates, the lower-energy rotational

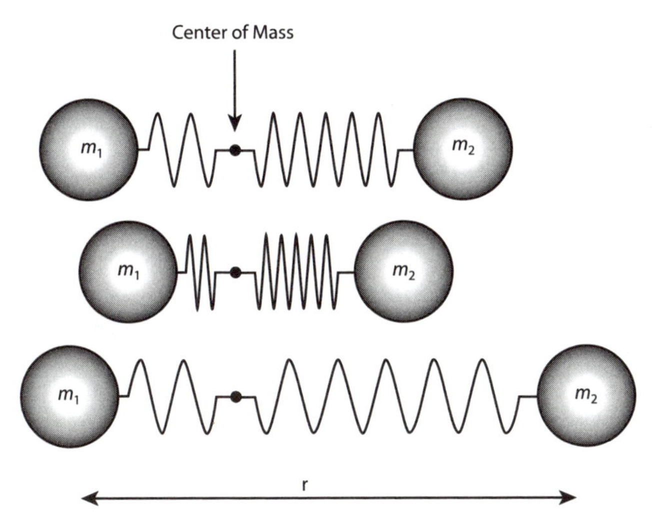

Figure 8.5 Schematic illustration of molecular vibration for a linear diatomic molecule. The simple harmonic oscillator is used as an approximate model for a diatomic molecule.

modes are also excited. Nevertheless, because the energy of the vibrational transition is much larger than that of the rotational transition, the vibrational energy levels and transitions can be considered independently of rotation to a good approximation. The so-called rotational-vibrational spectrum is an array of rotational lines grouped around a vibrational transition (Section 8.2.3).

To develop a conceptual understanding of vibrational energy levels, we will follow the discussion for rotational energy levels, again using a classical mechanics analogy limited to linear diatomic molecules. The classical analogy for diatomic molecular vibration is that of a simple harmonic oscillator. With this analogy we are limiting our attention to vibration along the axis joining the nuclei (and, again, ignoring rotational motion).

From Figure 8.3, we can see that the potential energy of a stable electronic state as a function of internuclear distance x can be approximated by a parabola. The parabola with the equilibrium point at the minimum of the potential energy means that as the atoms are displaced there is a restoring force pushing the atoms back to the equilibrium position. For a separation smaller than the equilibrium one, a strong repulsion develops. For separations greater than the equilibrium separation, the force becomes attractive. This is why the nuclei of the two atoms are essentially maintained at a more or less well-defined equilibrium separation. The approximate potential energy curve can be described by the potential energy for a simple harmonic oscillator,

$$V(x) = \frac{1}{2}Cx^2. \tag{8.20}$$

Here x is the distance from the center of mass. The value of the force constant C reflects the strength of the force holding the two atoms together.

The classical expression for the total energy of the diatomic molecule is

$$E_{\mathrm{v}} = \frac{1}{2}m_1 v_1^2 + \frac{1}{2}m_1 v_2^2 + V(x). \tag{8.21}$$

Using the reduced mass

$$\frac{1}{\mu} = \frac{1}{m_1} + \frac{1}{m_2} \tag{8.22}$$

and considering that the momenta of the nuclei, p, are equal (but of opposite magnitude), we have

$$E_{\mathrm{v}} = \frac{p^2}{2\mu} + V(x) = \frac{p^2}{2\mu} + \frac{1}{2}Cx^2. \tag{8.23}$$

This is the expression for the total energy for the diatomic molecule in terms of the reduced mass of the system,

We now associate the expression for the total energy with the quantum mechanical operator, $H\psi = E\psi$,

$$-\frac{\hbar^2}{2\mu}\frac{d^2\psi}{dx^2} + \frac{1}{2}Cx^2\psi = E_{\mathrm{v}}\psi. \tag{8.24}$$

The eigenvalue, or quantized energy level, solution to this equation is

$$E_{\mathrm{v}} = (\mathrm{v} + 1/2)h\left[\frac{1}{2\pi}\sqrt{\frac{C}{\mu}}\right], \tag{8.25}$$

which can also be written

$$\boxed{E_{\mathrm{v}_k} = h\nu_0(\mathrm{v}_k + 1/2) \qquad (\mathrm{v}_k = 0, 1, 2, ...),} \tag{8.26}$$

where we have associated $\frac{1}{2\pi}\sqrt{C/\mu}$ with the fundamental frequency ν_0. Here v is the vibrational quantum number. The subscript k denotes the normal modes ($k = 1, 2, 3, \ldots$). There is no degeneracy in vibrational energy states, that is,

$$g_{\mathrm{v}} = 1. \tag{8.27}$$

What is so interesting about using the simple harmonic oscillator approximation is that the derived energy levels are equally spaced with a separation of $h\nu_0$. This results from a vibrational transition selection rule $\Delta\mathrm{v} = \pm 1$ (Section 8.3.2). The equal energy level separation means in principle that the energy of a vibrational transition, and hence wavenumbers and frequencies, are the same for different transitions. In practice, anharmonicity makes the energy level separations different from each other.

We can estimate the energy of vibrational transitions using the association $\nu_0 = \frac{1}{2\pi}\sqrt{C/\mu}$ from our simple harmonic oscillator analogy, and with knowledge of C. If we take the force constant C for the CO molecule as 1860 N m^{-2} [4], we find $\nu_0 \sim 6.4 \times 10^{13}$ Hz, corresponding to an energy of about 4×10^{-20} J, a wavelength of about 4.3 μm, and a wavenumber of about 2100 cm^{-1}, in the infrared region of the electromagnetic spectrum. More commonly, the force constant C is determined from observations of a vibrational spectrum, rather than by taking the force constant to estimate the spectrum frequencies.

The lowest vibrational energy level is $E_0 = \frac{1}{2}h\nu_0$. That the zero-point energy is not actually zero is a quantum mechanical manifestation, in contrast to classical mechanics where the zero-point energy of a simple harmonic oscillator is zero. The zero-point energy represents the residual vibrational energy possessed by the molecule.

There are limitations to using the harmonic oscillator potential to describe molecular vibrational energy levels. We can see from Figure 8.3 that the potential energy curve for a stable electronic state is not actually a perfectly symmetric parabola. Away from the equilibrium position, as the internuclear separation decreases, the potential rises more steeply than as the internuclear separation increases. This sharp rise in potential energy comes from the repulsive forces as the nuclei approach each other. With increasing separation, the molecular bond is eventually broken. Thus the harmonic oscillator potential becomes unrealistic for both small and larger internuclear separations.

8.2.3 The Rotational-Vibrational Spectrum

The rotational-vibrational spectrum for a given vibrational transition $\Delta\mathrm{v}$ is divided into three "branches," the P-branch, Q-branch, and R-branch. These branches arise from the radiation selection rule that in a rotational transition $\Delta J = \pm 1$. The P-branch corresponds to $\Delta J = -1$ and the R-branch to $\Delta J = +1$. The Q-branch corresponds to no rotational transition. For diatomic molecules there are no transitions, resulting in a "gap" in the actual spectrum. See Figure 8.6. For an example, absorption cross-sections for the linear molecule CO are shown in Figure 8.7.

Molecules have specific vibrational modes, each with a fundamental frequency with its own set of quantum numbers. These are known as normal modes or fundamental vibrational modes. Each normal mode of vibration is independent of the others and corresponds to simultaneous vibrations of different parts of the molecule A linear molecule has only one fundamental vibrational mode, that is, only way that the molecule can vibrate (Figure 8.5). For linear molecules with more than two atoms, the number of normal modes of vibration is $3N - 5$, where N is the number of atoms and 5 is the degrees of freedom. For nonlinear molecules the number of normal modes of vibration is $3N - 6$. Figure 8.8 illustrates the different normal modes for a polyatomic linear molecule (CO_2) and a triatomic nonlinear molecule (H_2O).

In addition to the fundamental normal modes, molecules can have vibrational overtones. The first vibrational overtone would be $\mathrm{v} \rightarrow \mathrm{v} - 2$, the second vibrational overtone would be $\mathrm{v} \rightarrow \mathrm{v} - 3$, and so on. In the simple harmonic oscillator treatment of molecules, we have assumed that normal modes are fully independent. In reality, the normal modes can be coupled, giving rise to vibrations with combinations of the fundamental vibration frequencies, called combination bands.

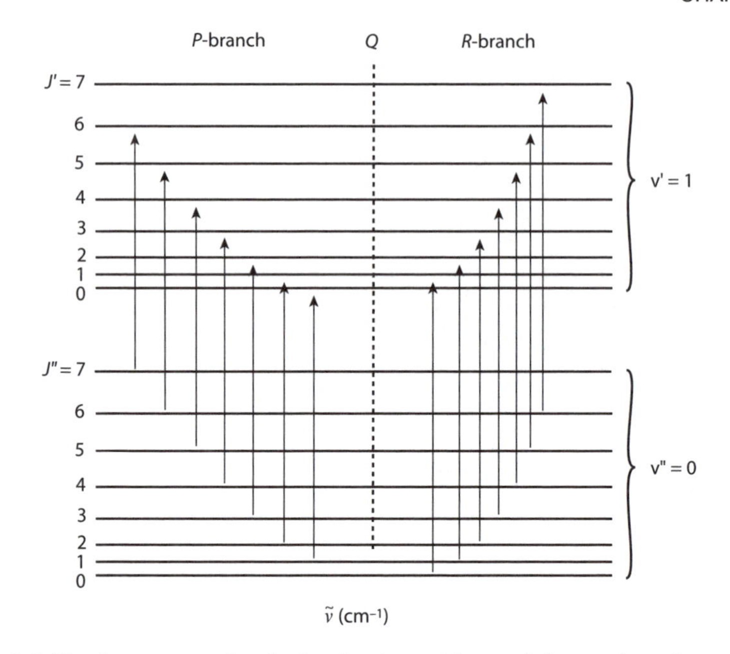

Figure 8.6 Simultaneous rotational vibrational transitions. $\Delta J = -1$ produces the P-branch, $\Delta J = -1$ produces the R-branch. For nonlinear molecules $\Delta J = 0$ produces the Q-branch.

8.2.3.1 Molecular Electronic Transitions

Molecules have electronic transitions that occur at much higher energy levels than vibrational or rotational transitions. We can see this by first considering how much less massive electrons are than the nuclei in the molecules, about a factor of 10^{-4} to 10^{-5}. By the Heisenberg uncertainty principle, the electrons travel much faster and must therefore have much higher energies than the rotational or vibrational motions of the nuclei. Due to this great energy difference, electronic transitions can be described independently of the rotational and vibrational transitions. The separation of electronic and nuclear motions is known as the Born-Oppenheimer approximation.

We can estimate the energy of electronic transitions by taking the momentum of an electron to be $\sim \hbar/a$, where $a \sim 1$ Å is a typical molecular size. Together with $E = p^2/m$ we have

$$E_{\mathrm{E}} \sim \frac{\hbar^2}{ma^2}. \tag{8.28}$$

The molecular electronic energy is a few eV (about a few $\times 10^{-19}$ J) or higher, which corresponds to photons at visible wavelengths or shorter.

Molecular electronic energy levels, like all atomic and molecular energy levels are quantized. Unlike for rotational and vibrational transitions, the electronic energy levels and transitions are too complex for us to present even a simple overview here. (See references listed at the end of this chapter for more information.) Elec-

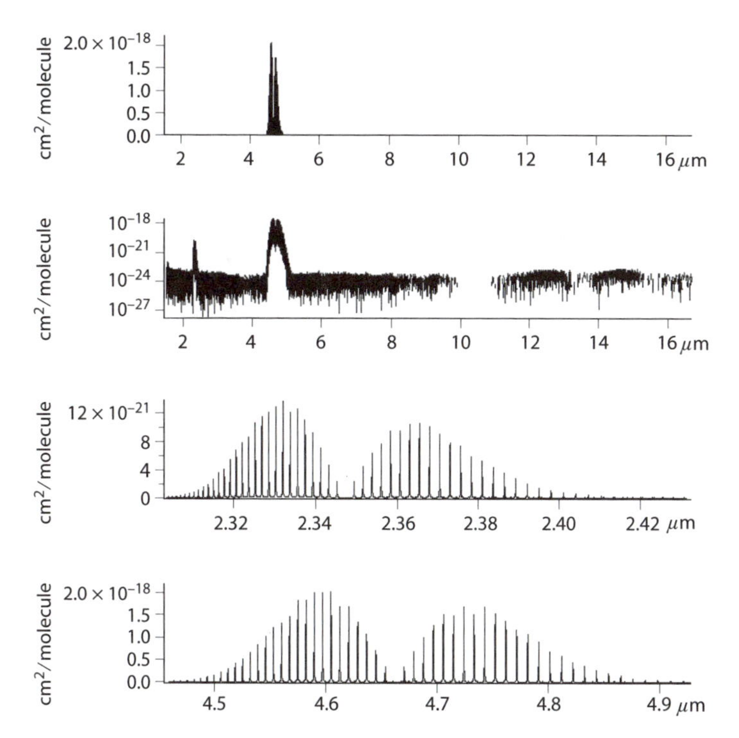

Figure 8.7 Simulated intensity of rotational-vibrational transitions of the CO molecule. The rotational transitions are centered around the P- and R-branches. The line intensities are approximately correct for a temperature of 300 K. The vibrational transition is the $\mathrm{v} = 1 - 0$ band. Adapted from [6].

tronic transitions can excite molecular vibrational and rotational transitions, resulting in an electronic band with finer structure from rotational and vibrational transitions.

It can be useful to have a handle on molecular notation. For linear molecules electronic states are designated by

$$^{(2S+1)}\Lambda^{(\pm)}_{(u,g)}, \tag{8.29}$$

where S is the electron spin quantum number, $(2S{+}1)$ is the degeneracy, (u, g) refers to odd or even wavefunction, and $\pm$ refers to the reflection symmetry of the wavefunction. The symbol Λ refers to the orbital wavefunction type (shape), and the first three levels are denoted as Σ, Π, Δ. An example of electronic energy levels and notation is given in Figure 8.9.

8.2.4 Energy Level Populations

We now turn from energy levels and their spacing to the number density of molecules in a given excited state, n_i, where i denotes an excited state. Often referred to as the energy level populations, or sometimes occupation numbers, the

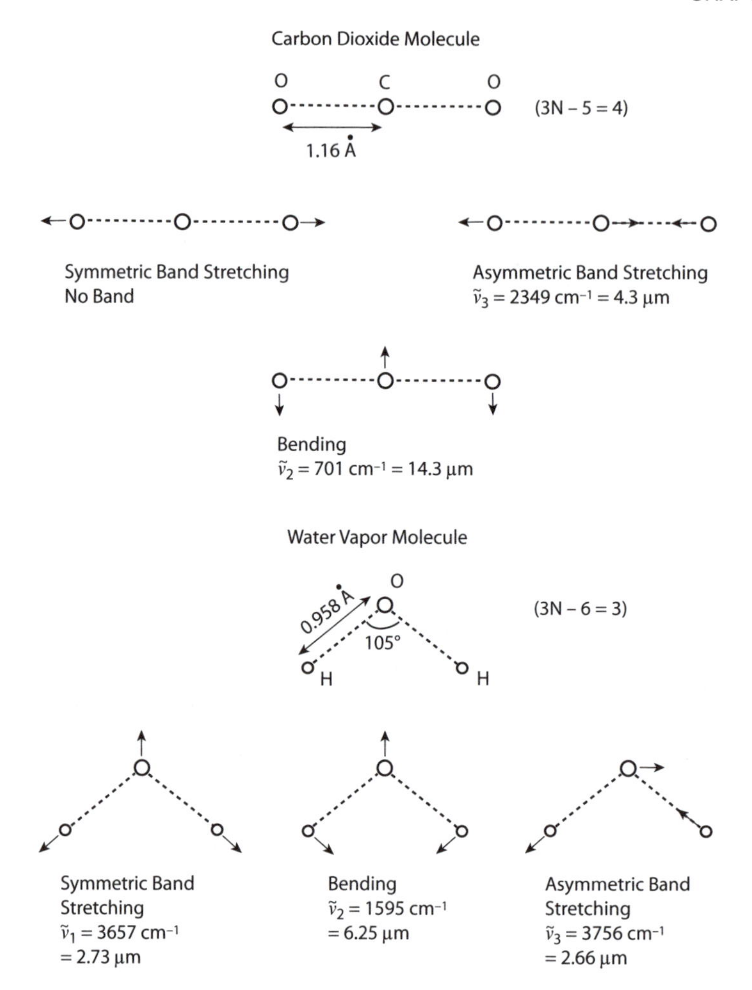

Figure 8.8 Schematic of molecular vibrational normal modes. The normal modes are indepenent of each other, corresponding to simultaneous vibrations of different parts of the molecule.

n_i are a basic component of the absorption coefficient (equation [8.1]). Under conditions of local thermodynamic equilibrium (LTE), we can prescribe a relatively simple formulation for the energy level populations. The formulation is simple because it depends only on temperature and not on the radiation field.

LTE is a local version of complete thermodynamic equilibrium. LTE is an approximation, whereby all precepts of thermodynamic equilibrium are assumed to hold, except that the radiation field departs from its Planckian value. LTE is a reasonable approximation valid in a local area of the atmosphere where any temperature, pressure, or chemical gradients are small compared to the photon mean free path. If collisions dominate, LTE implies a strict coupling of the matter component to the local radiation temperature, and can be used for regions of planetary

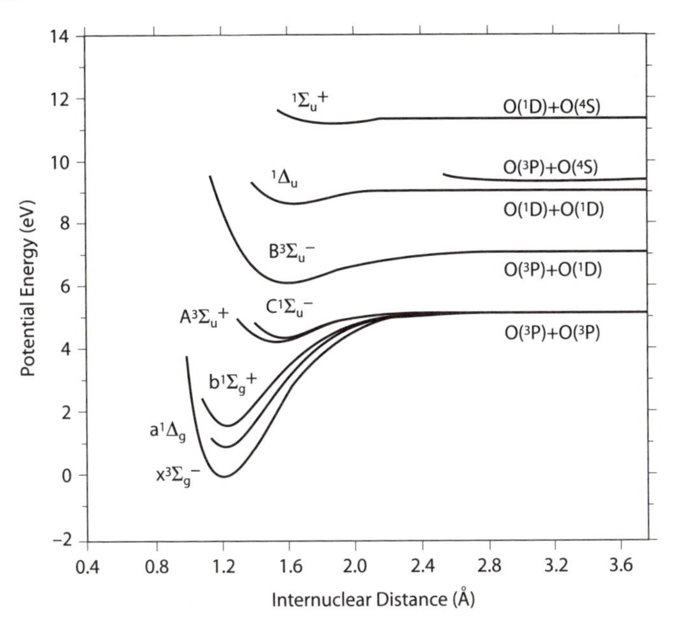

Figure 8.9 Simplified potential energy diagram for oxygen. The dissociation products are listed along the right side. Adapted from [7].

atmospheres where collisional processes dominate over radiative processes (see Figure 8.10). See Section 5.4 for an introduction to LTE and its major simplification to the radiative transfer problem.

Under LTE, the energy level populations for electronic, rotational, and vibrational states are determined by Boltzmann statistics. Hence the major simplication: energy level populations can be described by the Boltzmann formula

$$\frac{n_j}{n} = \frac{g_j \mathrm{e}^{-E_j/kT}}{\sum_i g_i \mathrm{e}^{-E_i/kT}}, \tag{8.30}$$

where $Q = \sum_i g_i \mathrm{e}^{E_i/kT}$ is the partition function. Here n is the total number density of a given molecule, n_j is the number density of the molecule with energy level j, E_j is the energy of level j from the ground state, k is Boltzmann's constant, and T is the temperature. Here g is the statistical weight of the level, also known as the energy degeneracy of the level. g is an integer. The degeneracy refers to the number of distinct states having the same energy E_j but with a different set of quantum numbers.

The Boltzmann distribution for molecular rotational states is

$$\frac{n_J}{n_0} = (2J+1)\mathrm{e}^{-BJ(J+1)/kT}, \tag{8.31}$$

where the values for g and E are from Section 8.2.2.1, and the subscript 0 refers to the ground state. For molecular vibrational states,

$$\frac{n_\mathrm{v}}{n_0} = \mathrm{e}^{-\mathrm{v}h\nu_0/kT}, \tag{8.32}$$

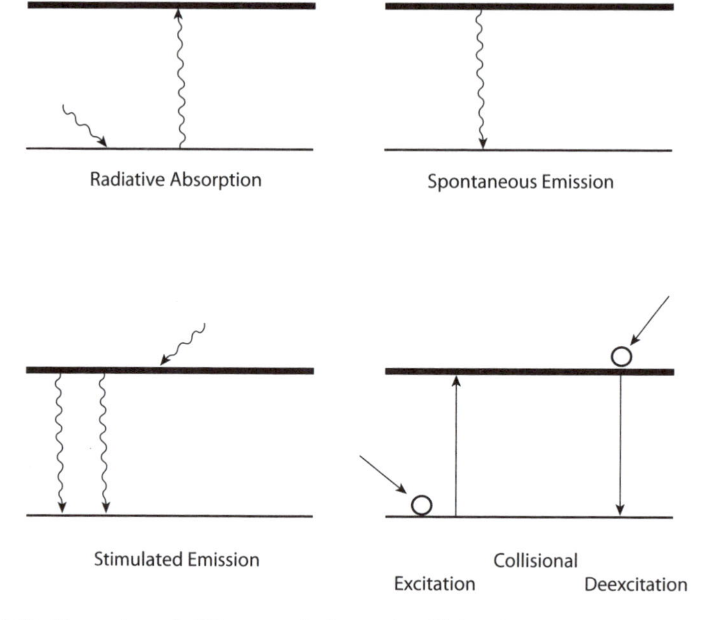

Figure 8.10 Illustration of different radiative and collisional processes for transitions between excited states, using a two-level atom for illustration. The wavy line indicates a photon. The circles represent atoms and the straight line indicates a collision. Collisions couple the matter and radiation temperature. Adapted from [8].

where the values for g and E are from Section 8.2.2.2. For a discussion of atomic energy level LTE distributions, see, for example, [9].

In planetary atmospheres LTE is valid where collisions couple the matter and radiation temperatures. In planetary atmospheres, LTE breaks down where collisions become less important than radiative transitions. One such case is at low pressures (low densities) where radiative processes dominate the competing collisional excitation processes (see Figure 8.10).

In the non-LTE case, how can the energy level populations be calculated? The answer is by use of the statistical equilibrium equations of detailed balance together with the radiation field. In detailed balance, every transition happens in both directions. In the case where radiative transitions dominate over collisional transitions, the equation of detailed balance is

$$n_u A_{ul} - n_u B_{ul} \overline{J}_\nu = n_l B_{lu} \overline{J}_\nu, \tag{8.33}$$

where the subscripts u and l refer to the upper and lower transitions, respectively. The variable $\overline{J}_\nu = \int_0^\infty J(\tau, \nu)\phi(\nu - \nu_0) d\nu$ is the mean radiation field over the transition.

The Einstein A and B coefficients in the above equation are as follows. A_{ul} in units of s^{-1} is the Einstein coefficient of spontaneous emission, B_{lu} in units of m^3 J^{-1} s^{-2} is the Einstein coefficient of induced absorption, and B_{ul} also in units of m^3 J^{-1} s^{-2} is the Einstein coefficient of stimulated emission. We leave it as an exercise to show that the Einstein coefficients are related to each other by

$$\begin{aligned} B_{ul} &= \frac{g_u}{g_l}\frac{c^2}{2h\nu^3}A_{ul}, \\ B_{ul} &= \frac{g_l}{g_u}B_{lu}, \\ A_{ul} &= -A_{lu}. \end{aligned} \tag{8.34}$$

The calculation of the energy level populations from detailed balance (equation [8.33]) requires the mean radiation field. The radiation field is derived from a solution to the radiative transfer equation. Yet, in the radiative transfer equation, the radiation field (via the intensity) depends in part on the energy level populations. In a solution of the radiative transfer equation under non-LTE conditions, each energy level of each atom or molecule would require an additional equation.

So far, non-LTE treatment is not usually warranted by the quality of exoplanet atmosphere data. One exception in the future might be high spectral resolution of transit transmission spectra. Transmission spectra probe high layers of the atmosphere where densities may be such that LTE breaks down. In contrast, in Earth's atmosphere (Figure 9.2), LTE holds for the lower atmosphere below altitudes of about 60–70 km. A thorough description of non-LTE conditions of validity and treatment can be found in [9, 10].

8.3 MOLECULAR ABSORPTION CROSS SECTIONS

We now move on to the absorption cross section $\sigma_{lu}(\nu)$ or $\sigma_{lu}(\tilde{\nu})$ for a molecular rotational or vibrational energy transition. The absorption cross section is made up of two components, the strength of the transition itself and a line broadening function. Molecular lines are never truly monochromatic, but due to a variety of physical processes are spread out in frequency. We therefore define the cross section in terms of the transition line strength S_{lu} and the normalized line broadening function $f(\tilde{\nu} - \tilde{\nu}_0)$,

$$\sigma_{lu} = S_{lu} f(\tilde{\nu} - \tilde{\nu}_0). \tag{8.35}$$

S is more formally called the spectral line intensity in units of cm^{-1}/(molecule cm^{-2}) and $\tilde{\nu}_0$ is the wavenumber of a monochromatic line. f is normalized as

$$\int_{-\infty}^{\infty} f(\tilde{\nu} - \tilde{\nu}_0) d\tilde{\nu} = 1. \tag{8.36}$$

We may also write

$$\int_{-\infty}^{\infty} \sigma_{lu}(\tilde{\nu}) d\tilde{\nu} = S_{lu}, \tag{8.37}$$

again stating that the absorption coefficient is not delta-function-like at a monochromatic frequency, but has a spread in frequency owing to line broadening mechanisms.

8.3.1 Line Strengths

Let us treat the line strength by first relating it to the Einstein coefficients. We have the definition (equation [8.35])

$$\alpha_l = n_l \sigma_{lu} = n_l S_{lu} f(\tilde{\nu} - \tilde{\nu_0}). \tag{8.38}$$

But we could also show that the energy removed from a beam of radiation is

$$\alpha_l = \frac{h\tilde{\nu}_0}{4\pi c}\left(n_l B_{lu} - n_u B_{ul}\right) f(\tilde{\nu} - \tilde{\nu}_0), \tag{8.39}$$

where we have treated stimulated emission as negative absorption. We can write

$$\alpha_l = n\left[\frac{h\tilde{\nu}_0}{4\pi c}\frac{n_l}{n} B_{lu}\left(1 - \mathrm{e}^{-h\tilde{\nu}_0/ckT}\right)\right] f(\tilde{\nu} - \tilde{\nu}_0), \tag{8.40}$$

where we have assumed LTE to use the Boltzmann distribution equation for energy level populations (equation [8.30]). We associate the LTE line strength with the term in square brackets,

$$S_{lu} = \left[\frac{h\tilde{\nu}_0}{4\pi c}\frac{n_l}{n} B_{lu}\left(1 - \mathrm{e}^{-h\tilde{\nu}_0/ckT}\right)\right], \tag{8.41}$$

and we have implicitly redefined $\alpha = n\sigma_{lu}$, with σ_{lu} weighted by n_l/n.

We are now left to describe where the Einstein B coefficient comes from. B_{lu} comes from quantum mechanics. Essentially,

$$B_{lu} = \frac{8\pi^3\tilde{\nu}_0}{3hc}\left|\int \Phi_u^*(\overline{\mu}\Phi_l)dV\right|^2 \left|\int \Sigma_u^*\Sigma_l d\sigma\right|^2. \tag{8.42}$$

Here $\Phi(r, \theta, \phi)$ is the space-dependent part of the wavefunction, $\Sigma(\sigma)$ is the spin-dependent part of the wavefunction, and $\overline{\mu}$ is the dipole moment operator. From all of the above, if we were able to carry out detailed quantum mechanical calculations, we could compute the spectral line intensity of a given molecular transition.

The theoretical computation or laboratory measurement of lines and line transition identification is a whole research field of its own. In exoplanet atmosphere research we typically take the line strength directly from a molecular database, such as HITRAN [11]. In the HITRAN database, the line strength at STP is given, and one may scale it according to

$$\boxed{S_{lu}(T) = S_{lu}(T_{\mathrm{ref}})\frac{Q(T_{\mathrm{ref}})}{Q(T)}\frac{\exp(-c_2 E_l/T)}{\exp(-c_2 E_l/T_{\mathrm{ref}})}\frac{[1 - \exp(-c_2\tilde{\nu}_{lu}/T)]}{[1 - \exp(-c_2\tilde{\nu}_{lu}/T_{\mathrm{ref}})]}}. \tag{8.43}$$

Here E_l is the lower state energy in cm^{-1}, c_2 is the second radiation constant hc/k = 1.4388 cm K, and $\tilde{\nu}$ is still the wavenumber in cm^{-1}. The second term on the right-hand side accounts for a temperature scaling of the partition function, the third term for the ratio of the Boltzmann populations, and the fourth term for stimulated emission.

8.3.2 Selection Rules for Molecules

Out of all the existing energy levels of a given molecule (including electronic, rotational, and/or vibrational energy levels), transitions happen only between some of the energy levels. The allowed transitions are described by "selection rules."

A selection rule is a quantum mechanical rule describing transitions that are permitted. The transition selection rules ultimately come from equation [8.42], from a quantum mechanical description of the eigenfunction or the wavefunctions of each of the two energy levels involved in a given energy transition. Where equation [8.42] vanishes, there is no transition. Where $B_{lu} \gg 0$, the transition is "perimtted," and where B_{lu} is very small, the transition is "forbidden." A forbidden transition is forbidden to "first order" only; they are rare and hence weak. We can think of forbidden as meaning unfavorable.

For example, if there is no dipole moment, equation [8.42] goes to zero. This is a more formal basis for our earlier conceptual remarks that a molecule must posses a dipole moment in order to interact with electromagnetic radiation to produce a rotational or rotational-vibrational spectrum.

As a second example, $\Phi_u^*(\overline{\mu}\Phi_l)$ must be an even function to prevent significant cancellation between volume elements. Since $\overline{\mu}$ is an odd function in space, then Φ_u^* and Φ_l must have opposite parity. This formally leads to the rotational and vibrational energy transition selection rules as follows. For rotational energy transitions in a linear molecule $\Delta J = \pm 1$. Additionally, for a nondiatomic simultaneous rotational-vibrational transition where the vibration is perpendicular to the rotational axis, $\Delta J = 0, \pm 1$. For vibrational transitions in all molecules, $\Delta \text{v} = \pm 1$. This means in principle that, for polyatomic molecules, overtones (e.g., $\text{v} = 0 \rightarrow \text{v} = 2$) are not allowed, nor are combination bands (e.g., transitions in two normal modes at the same time). In reality, the selection rules can be violated if perturbations affect the symmetry of the wavefunctions, namely, overtone and combination bands are observed.

8.3.3 Line Broadening

We have one remaining ingredient to describe for the absorption cross section and that is line broadening. Molecular energy transitions are not precise, that is, they do not absorb or emit photons that are monochromatic. Monochromatic emission is practically never observed. Energy levels during energy transitions are always slightly altered for an individual molecule, by both internal and external influences. These changes in energy cause spectral lines to have finite widths in frequency.

8.3.3.1 Natural Broadening

Natural broadening is line broadening caused by the uncertainty in the values of energy levels. If the energy levels are not precisely known, the energy and hence frequency of an emergent photon will be different at different times. Even though

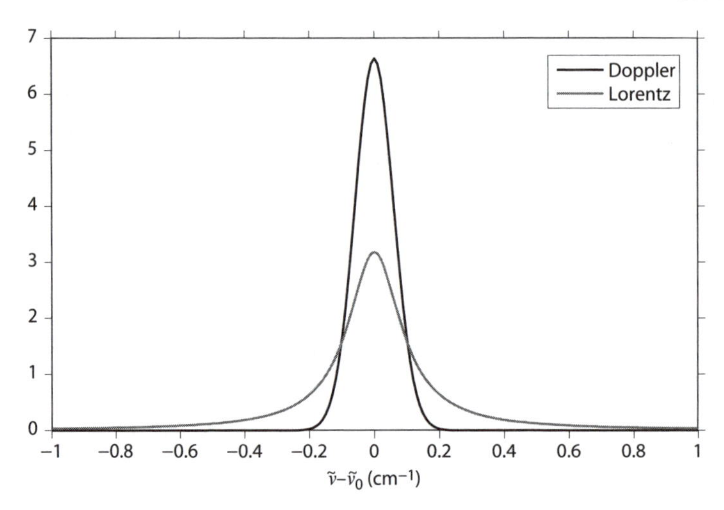

Figure 8.11 Comparison of Doppler and collisional broadening. The y-axis is the normalized probability density.

natural broadening is negligible in planetary atmospheres we describe it here for completeness.

Natural broadening comes from the Heisenberg uncertainty principle, that time and energy cannot both be precisely known,

$$\Delta E \Delta t = \frac{h}{2\pi} \tag{8.44}$$

or

$$\Delta\nu\Delta t = \frac{1}{2\pi}. \tag{8.45}$$

We may then write the line broadening in frequency as

$$\Delta\nu = \frac{1}{2\pi\Delta t} = 2A_{ul}, \tag{8.46}$$

where A_{ul} is the Einstein A coefficient described in equation [8.34].

Natural broadening is described by the Lorentz profile

$$f_{\mathrm{N}}(\nu - \nu_0) = \frac{1}{\pi}\frac{\gamma_{\mathrm{N}}}{(\nu-\nu_0)^2 + \gamma^2}, \tag{8.47}$$

where γ_{N} is the full-width at half maximum (FWHM), $\gamma \simeq 1/2\pi t$, where t is the radiative lifetime. Note that α is often used instead of γ; here we are trying to avoid confusion with the absorption coefficient α.

8.3.3.2 Doppler Broadening

Doppler broadening arises from the difference in thermal velocities of atoms and molecules. Recall the Doppler shift

$$\nu = \nu_0\left(1 \pm \frac{v}{c}\right), \tag{8.48}$$

where ν is the frequency, ν_0 is the rest frequency, v is the velocity, and c is the speed of light.

There is a distribution of speeds along the line of sight which creates line broadening. We will assume that the distribution of speeds is given by the Maxwell-Boltzmann distribution

$$P(v)dv = \sqrt{\frac{m}{2\pi kT}} \exp\left(\frac{-mv^2}{2kT}\right) dv, \tag{8.49}$$

where $P(v)dv$ is the fraction of particles with speeds v along the line of sight.

We can substitute the Doppler shift (equation [8.48]) into the Maxwell-Boltzmann distribution to find an expression for Doppler line broadening,

$$\boxed{f_{\mathrm{D}}(\nu - \nu_0) = \frac{1}{\gamma_{\mathrm{D}}\sqrt{(\pi)}} \exp\left[-\left(\frac{\nu - \nu_0}{\gamma_{\mathrm{D}}}\right)\right],} \tag{8.50}$$

where

$$\gamma_{\mathrm{D}} = \nu_0\sqrt{\frac{2kT}{mc^2}}. \tag{8.51}$$

Because the Doppler broadening equation is a Gaussian, γ_{D} is the standard deviation. The FWHM of the Doppler-broadened line is

$$\mathrm{FWHM} = \gamma_{\mathrm{D}}\sqrt{\ln 2}. \tag{8.52}$$

We can see from the exponential term in the Doppler broadening equation (equation [8.50]) that the spectral line will be more intense at line center and weaker in the line wings (see Figure 8.11). From the Doppler broadening equation we can also see how the line broadening depends on temperature.

8.3.3.3 Collisional Broadening

Collisional, or pressure, broadening describes line broadening due to collisions between atoms and molecules. If we take atomic and molecular collisions to be instantaneous, the principal effect of the collision is to destroy the phase coherence of the emitted wave train. In other words, after the collision, the molecule starts emitting at another phase; new phases are randomly distributed. This random distribution of phases together with the Poisson distribution for collisions leads to the Lorentz profile

$$\boxed{f_{\mathrm{L}}(\nu - \nu_0) = \frac{1}{\pi}\frac{\gamma}{(\nu - \nu_0)^2 + \gamma^2}.} \tag{8.53}$$

Here ν_0 is the frequency of an ideal monochromatic line, and γ is the half width of the line at half maximum. ν and ν_0 may be defined as wavenumber as long as γ is also defined in the same units. For a derivation of the Lorentz profile for both collisional broadening and natural line broadening, see [7,8].

The parameter γ is complicated to calculate, and comes from quantum mechanics calculations or from laboratory measurements. Based on the kinetic theory of gases, however, we may scale γ as

$$\gamma = \gamma_0 \frac{P}{P_0} \left(\frac{T_0}{T} \right)^n , \tag{8.54}$$

where the subscript 0 refers to standard temperature ($T_0 = 273$ K) and pressure ($P_0 = 1013$ mbar). The broadening exponent n has a classical value that is 0.5, whereas real molecular lines have broadening exponents ranging from about 0.4 to 0.75. In the above equation, increased pressure can be understood as increasing the concentration of molecules available for collisions, and hence increasing γ. We can think of an increase in temperature as an increase in velocity whereby the molecules spend less time perturbing each other because they are in each other's vicinity for a shorter time. Finally, if an approximate value of the collisional broadening parameter is warranted, the van der Waals collisional broadening theory can be used to estimate γ [e.g., 9].

8.3.3.4 Combined Doppler and Lorentz Line Broadening

What is the form of a combined Doppler and Lorentz broadening profile? To account for all possible thermal velocities, a convolution of the two line shapes is needed,

$$f_{\mathrm{Voigt}}(\nu - \nu_0) = \int_{-\infty}^{\infty} f_{\mathrm{L}}(\nu' - \nu_0) f_{\mathrm{D}}(\nu - \nu') d\nu' . \tag{8.55}$$

Substituting the Lorentz and Doppler broadening terms we have

$$f_{\mathrm{Voigt}}(\nu - \nu_0) = \frac{1}{\pi^{3/2}} \frac{\gamma}{\gamma_{\mathrm{D}}} \int_{-\infty}^{\infty} \frac{1}{(\nu' - \nu_0)^2 + \gamma^2} \exp\left[-\frac{(\nu - \nu')^2}{\gamma_{\mathrm{D}}^2} \right] d\nu' . \tag{8.56}$$

We can simplify the above expression by defining

$$t \equiv \frac{(\nu - \nu')}{\gamma_{\mathrm{D}}}; \quad y \equiv \frac{\gamma}{\gamma_{\mathrm{D}}}; \quad x \equiv \frac{(\nu - \nu_0)}{\gamma_{\mathrm{D}}} . \tag{8.57}$$

We then find

$$\boxed{f_{\mathrm{Voigt}}(\nu - \nu_0) = \frac{1}{\gamma_{\mathrm{D}} \sqrt{\pi}} K(x, y),} \tag{8.58}$$

where

$$K(x, y) = \frac{y}{\pi} \int_{-\infty}^{\infty} \frac{1}{y^2 + (x - t)^2 \exp(-t^2)} dt . \tag{8.59}$$

8.4 RAYLEIGH SCATTERING

Rayleigh scattering is the dominant opacity at short wavelengths and because of its λ^{-4} dependence is responsible for Earth's blue sky (see Figures 1.2 and 6.11).

Rayleigh scattering is derived and described in detail in many textbooks, e.g., [10]. Here we simply provide formulae for completeness. Rayleigh scattering is valid for molecules or for any particle that is much smaller than the wavelength of light.

The Rayleigh scattering cross section in m^2 is

$$\sigma_{\rm scat}(\lambda) = \alpha_{\rm p}^2 \frac{128\pi^5}{3\lambda^4}, \tag{8.60}$$

where $\alpha_{\rm p}$ is the polarizability

$$\alpha_{\rm p} = \frac{3}{4\pi n}\left(\frac{m^2-1}{m^2+2}\right), \tag{8.61}$$

where n is the number density of molecules and m is the refractive index (or a weighted average of the refractive indices). Using the definition of polarizability, and considering that $m \sim 1$, the Rayleigh scattering cross section can be written

$$\sigma_{\rm scat}(\lambda) = \frac{8\pi^3\left(m^2-1\right)^2}{3\lambda^4 n^2}. \tag{8.62}$$

A useful numerical formula for the Rayleigh scattering cross section for Earth's atmosphere is

$$\sigma_{\rm scat}(\lambda) = \lambda^{-4}\sum_{i=0}^{3} a_i \lambda^{-2i} \times 10^{-28}, \tag{8.63}$$

in units of cm^2. This formula is accurate to 0.3% on the range $0.205 < \lambda < 1.05$ μm. Here the coefficients are a_0 = 3.9729066, a_1 = 4.6547659, a_2 = 4.5055995 $\times 10^{-4}$ and a_3 = 2.3229848 $\times 10^{-5}$ [12].

For H_2 a convenient formula can be found in [13],

$$\sigma_{\rm scat}(\lambda) = \frac{8.14\times 10^{-13}}{\lambda^4} + \frac{1.28\times 10^{-6}}{\lambda^6} + \frac{1.61}{\lambda^8}, \tag{8.64}$$

in units of $Å^2$.

8.5 CONDENSATE OPACITIES

We now turn to discuss the absorption and scattering coefficients of condensates. Condensates are solid particles, including not only liquid water droplets and water ice that make up Earth atmospheric clouds, but any cloud or haze particle that is liquid or solid. We must treat condensates separately from gas particles because as aggregrates of molecules they have no well-defined energy levels or rotational or vibrational states.

We can see many beautiful examples of scattering by condensates. Bright colorful single and double rainbows arise from refraction of sunlight through spherical water droplets. Another, more rare example is the so-called glory. The glory appears as circular colored rings inside which is the observer's shadow. The glory arises from light backscattered toward its source by water droplets of uniform size. The glory is commonly observed from airplanes or mountain tops, since the observer has to be directly between the water droplets and the Sun. The rainbow, glory, and other scattering phenomena are caused by reflection, refraction, and diffraction of light by or through particles.

8.5.1 Analytic Phase Functions

Condensate scattering is described using the single scattering phase function, because of the strong directional dependence of condensate scattering. The phase function describes the redirection in the incident intensity to the outgoing intensity and as such the 3D directional scattering probability. The phase function is described in terms of the scattering angle Θ and is normalized:

$$\frac{1}{4\pi}\int_{\Omega} P(\Theta) d\Omega = 1. \tag{8.65}$$

Θ can be expanded as

$$\begin{aligned}\cos\Theta &= \Omega\cdot\Omega' = \Omega_x\Omega'_x + \Omega_y\Omega'_y + \Omega_z\Omega'_z \\ &= \cos\theta\cos\theta' + \sin\theta\sin\theta'\cos(\phi' - \phi)\end{aligned} \tag{8.66}$$

in the Cartesian and spherical polar coordinate system described in Figure 2.4, but using θ as the angle away from the z-axis instead of ϑ. Here the prime refers to the direction of incidence and the terms with no prime refer to the direction after scattering. The term forward scattering is used for $\Theta < \pi/2$ and backward scattering for $\Theta > \pi/2$.

Some analytic examples of the phase function include isotropic scattering

$$P(\Theta) = 1; \tag{8.67}$$

Rayleigh scattering

$$P(\Theta) = \frac{3}{4}(1 + \cos^2\Theta); \tag{8.68}$$

and the Henyey-Greenstein phase function

$$P(\Theta) = \frac{1 - g^2}{(1 + g^2 - 2g\cos\Theta)^{3/2}}. \tag{8.69}$$

The Henyey-Greenstein phase function is a one-parameter fit to an actual phase function, but actually has no physical basis. The variable g is the anisotropy parameter, $-1 \le g \le 1$, and $g = 0$ for isotropic scattering or symmetric scattering,

$$g = \langle\cos\Theta\rangle = \frac{1}{4\pi}\int_{\Omega}\cos\Theta P(\Theta) d\Omega. \tag{8.70}$$

The Henyey-Greenstein phase function has been widely used because of its simplicity and convenient form for numerical simulations. Other phase functions can be described by an expansion in terms of the Legendre polynomials (see, e.g., [8, 10]). For many cases of condensate scattering, there is no analytic form, and we resort to tables generated by geometric optics or by Mie theory.

8.5.2 Description of Phase Functions from Geometric Optics

Geometric optics provides us with a conceptual description of phase function features. The geometric optics approximation to light scattering is valid when the scattering particle is very large compared to the wavelength of light. By considering a number of uniformly spaced rays striking a particle (Figure 8.12), taking

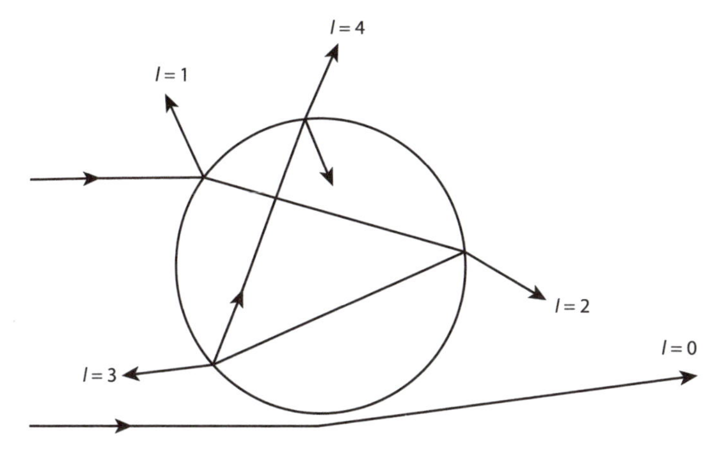

Figure 8.12 Paths of light rays reflected, refracted, or diffracted by a sphere. $l = 0$ shows a diffracted ray, $l = 2$ shows a twice refracted ray, $l = 3$ shows a ray with one internal reflection, and $l = 4$ shows a ray with two internal reflections. Adapted from [14].

into account reflection, transmission, and refraction, and summing the results over all incident rays and all significant components ($l = 0, 1, 2, \ldots$), one can derive a phase function like that shown in Figure 8.13.

Following [14] we will describe the phase function features using the indices of refraction of liquid water. Let us go through some of the phase function features in Figure 8.13, which compares results from geometric optics to Mie theory for different sized particles. All of the phase functions in Figure 8.13 notably show very similar features. At zero degrees, in the forward scattering direction, there is a very strong feature due to diffraction. We can see that diffraction increases as the particle size increases compared to the wavelength of light. The next notable features are at 137° and 130°. These features are the primary and secondary rainbows, related to rays with $l \geq 3$. At the backscattering angle (near 180°) there is a huge effect for large particles—this is the glory. The glory feature is not due to reflected rays ($l = 0$) but to surface waves on the sphere originating by interference. We can see from Figure 8.13 that features caused by light interference are not captured by geometric optics.

8.5.3 Mie Theory

We now turn to the Mie theory, which describes how to derive the phase functions and absorption and scattering cross sections of solid or liquid particles. Mie theory is a solution to Maxwell's equations for scattering of electromagnetic radiation of spherical particles. Mie theory is actually an analytic solution to Maxwell's equations and not a physical theory. It is fascinating that the Mie theory was developed over 100 years ago by Mie, and independently by others. In the limit $r \gg \lambda$ Mie

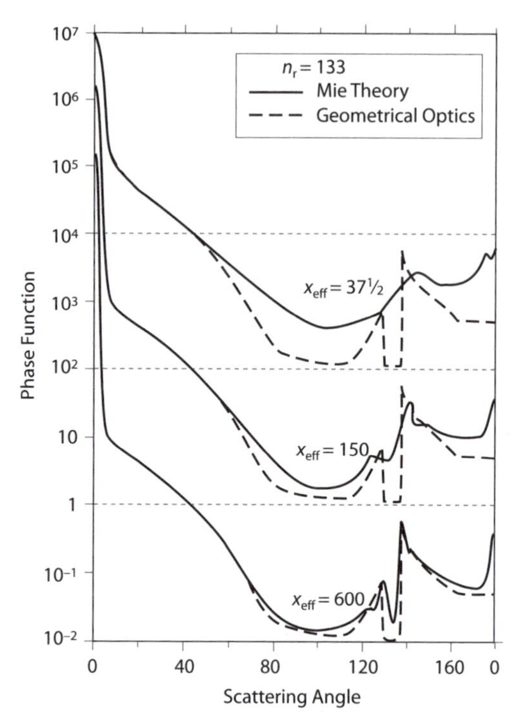

Figure 8.13 The phase function for single scattering by spherical liquid water droplets. A comparison of geometric optics and Mie theory phase functions. Mie theory captures interference between light rays whereas geometric optics does not. The phase functions for different size parameters $x = 2\pi r/\lambda$ are shown, where the subscript refers to an effective size for a particle distribution. Adapted from [14].

theory agrees with geometric optics and in the limit $r \ll \lambda$ Mie theory reduces to Rayleigh scattering. Here r is the particle radius and λ is wavelength of light as before. Mie theory is derived and applied in great detail in several excellent references [8, 10, 14–16]. Here we summarize the most important formulae, following [8, 14], to try to capture how the calculation would be carried out.

The main parameters we are interested in are those we need for the radiative transfer equation. The parameters include the scattering phase function $P(\Theta)$ (from which the related scattering asymmetry factor g can be computed), the scattering cross section $\sigma_{\rm scat}$, and the absorption cross section $\sigma_{\rm abs}$. In Mie theory the absorption and scattering cross sections are written in terms of the scattering and absorption effieniencies $Q_{\rm scat}$ and $Q_{\rm abs}$, $Q_{\rm scat} = \frac{\sigma_{\rm scat}}{\pi r^2}$ where r is the particle radius.

To get a handle on the origin of the above parameters of interest, we must take the definition of a scattering cross section as the amount of energy removed from the direction of incidence due to a single scattering event. The scatterer is the center of a sphere with radius R whereby the energy is redistributed isotropically on the surface of the sphere. The scattering cross section can be related to the intensity before and after scattering

$$I(\Theta) = I(0)\frac{\sigma_{\text{scat}}}{R^2}\frac{P(\Theta)}{4\pi} \tag{8.71}$$

Here and in the following discussions, I is the scattered radiation only. We therefore need to understand how the scattered radiation is calculated from the incident radiation.

The electric field of scattered radiation at a distance R (in the far field defined by $R \gg \lambda$) from a spherical particle can be represented by

$$\left\{ \begin{array}{c} E_\perp^s \\ E_\parallel^s \end{array} \right\} = \frac{\exp(-ikR + ikz)}{ikR} \left\{ \begin{array}{cc} S_1(\Theta, \phi) & 0 \\ 0 & S_2(\Theta, \phi) \end{array} \right\} \left\{ \begin{array}{c} E_\perp^i \\ E_\parallel^i \end{array} \right\}. \tag{8.72}$$

Here Θ is the scattering angle, and ϕ is an azimuthal angle measured about the direction of scatter. The incident radiation is propagating in the z-direction. For nonspherical particles, the above 2×2 scattering matrix is full. Here $E_\perp$ and $E_\parallel$ are the components perpendicular and parallel to the plane of scattering, where the plane of scattering is defined as the plane containing the directions of incidence and scattering. The superscripts i and s refer to incident and scattered, respectively. Our discussion is valid for isotropic, homogeneous spheres that scatter independently of each other (interference of light scattered by different particles is negligible).

To relate the above description of the electric field to the intensity before and after scattering we will use the definition in Section 7.2.2 and equation [7.7].

$$I = \langle E_\parallel E_\parallel^* + E_\perp E_\perp^* \rangle = \langle A_\parallel^2 + A_\perp^2 \rangle, \tag{8.73}$$

so that using equation [8.72], we then have

$$I = \frac{\sigma_{\text{scat}}}{4\pi} P^{11} I_0, \tag{8.74}$$

where σ_{scat} is the scattering cross section and

$$\frac{\sigma_{\text{scat}}}{4\pi} P^{11} = I/I_0 = \frac{1}{2}(S_1 S_1^* + S_2 S_2^*). \tag{8.75}$$

For the other P matrix elements see [1].

Mie theory also gives the scattering and extinction cross sections as

$$\sigma_{\text{scat}} = \pi r^2 \frac{2}{x^2} \sum_{n=1}^{\infty} (2n+1)(a_n a_n^* + b_n b_n^*), \tag{8.76}$$

$$\sigma_{\text{ext}} = \pi r^2 \frac{2}{x^2} \sum_{n=1}^{\infty} (2n+1)\mathcal{R}(a_n + b_n), \tag{8.77}$$

where $\mathcal{R}$ denotes the real part of a_n and b_n, $\sigma_{\text{abs}} = \sigma_{\text{ext}} - \sigma_{\text{scat}}$, and $\tilde{\omega} = \sigma_{\text{scat}}/\sigma_{\text{ext}}$. Here we have used the size parameter $x \equiv 2\pi r/\lambda$.

The heart of the Mie theory is the solution for the scattering amplitudes S_1 and S_2 and their parameters:

$$S_1 = \sum_{n=1}^{\infty} \frac{2n+1}{n(n+1)} [a_n \pi_n + b_n \tau_n], \tag{8.78}$$

$$S_2 = \sum_{n=1}^{\infty} \frac{2n+1}{n(n+1)} [b_n \pi_n + a_n \tau_n]. \tag{8.79}$$

The Mie theory solutions for S_1 and S_2 are properly interpreted as a multipole expansion of the scattered radiation. For example, a_1, a_2, and a_3 are the amount of electric dipole, quadrupole, and octopole radiation. The parameters b_n are similar coefficients for magnetic multipole radiation. Both a_n and b_n are generally complex. The computation of a_n and b_n is the central problem in computing the particle absorption and scattering cross sections and the phase functions. Their computation is related to spherical Bessel functions, and they can also be computed from recursion relations. From [8],

$$a_n = \frac{m\phi_n(mx)\phi_n'(x) - \phi_n(x)\phi_n'(mx)}{m\phi_n(mx)\zeta_n'(x) - \zeta_n(x)\phi_n'(mx)}, \tag{8.80}$$

$$b_n = \frac{\phi_n(mx)\phi_n'(x) - m\phi_n(x)\phi_n'(mx)}{\phi_n(mx)\zeta_n'(x) - m\zeta_n(x)\phi_n'(mx)}, \tag{8.81}$$

where $m = m_{\mathrm{p}}/m_{\mathrm{a}}$ is the ratio of the refractive indices of the particle (p) and the surrounding atmosphere (a). The index of refraction is complex, $m_{\mathrm{p}} = n_{\mathrm{p},r} + in_{\mathrm{p},i}$, where r refers to the real part of the refractive index and i to the complex part. The Ricatti-Bessel functions are ϕ and ζ, with primes denoting differentiation with respect to x or mx.

The parameters π and τ are functions only of the scattering angle Θ, are related to the Legendre polynomials, and are computed from recursion relations:

$$\pi_n(\Theta) = \frac{P_n^1(\Theta)}{\sin\Theta}, \tag{8.82}$$

$$\tau_n(\Theta) = \frac{dP_n^1(\Theta)}{d\Theta}, \tag{8.83}$$

where P_n is the associated Legendre polynomial. The first few terms are $\pi_1(\Theta) = 1$, $\pi_2(\Theta) = 3\cos(\Theta)$, $\tau_1(\Theta) = \cos(\Theta)$, $\tau_2(\Theta) = 3\cos 2\Theta$.

We now finish the Mie theory section with a general discussion of significant points to take away. We emphasize that the expressions for a_n and b_n are related only to the index of refraction and the size parameter $x \equiv 2\pi r/\lambda$, where again r is the particle radius and λ is the wavelength of light. In addition, the π_n and τ_n are functions only of the scattering angle Θ.

The solutions to S_1 and S_2 are infinite series. How many terms should actually be included? This depends on the size parameter x, since a_n and b_n both rapidly approach zero when $n \gg x$. The total number of terms need be only slightly larger than x.

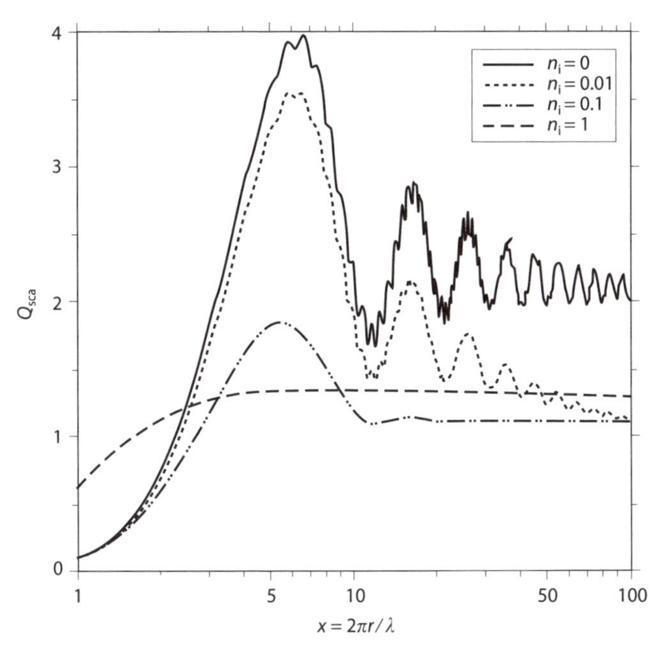

Figure 8.14 The scattering efficiency factor $Q_{\rm scat}$ as a function of the size parameter $x = 2\pi r/\lambda$. The refractive index $n_r = 1.33$ is that for liquid water. Different cases of the imaginary refractive index are shown and for decreasing values the interference effects diminish. Adapted from [14].

Mie theory is valid for spherical particles only. For nonspherical particles geometric optics approximations must be used.

Let us proceed to describe the outcomes of computations of Mie theory [14]. We will start with the scattering efficiency as a function of size parameter $x = 2\pi r/\lambda$. Figure 8.14 shows the scattering efficiency parameter $Q_{\rm scat} = \frac{\sigma_{\rm scat}}{\pi r^2}$ with a fixed real refractive index $n_r = 1.33$, but for varying complex refractive indices. Note that this is the refractive index for liquid water, and the complex part of the refractive index of water is negligible. The real part of the refractive index indicates scattering, whereas the imaginary part indicates the amount of absorption. If $n_i = 0$, the condensate particle does not absorb, it only scatters radiation. $Q_{\rm scat}$ as a function of the size parameter x shows ripples in the curve that are most prominent for $n_i = 0$. The peaks of enhanced scattering are caused by interference effects between diffracted light rays and rays that refract twice through the particle. This effect gets damped out for absorptive particles (e.g., increasing n_i). Real clouds are made of a distribution of particle sizes (see Section 4.4.4). The ripple effect also gets washed out for a dispersion of particle sizes (Figure 8.15).

Mie theory results for three different high-temperature condensates relevant for hot Jupiters are shown in Figures 8.16 and 8.17.

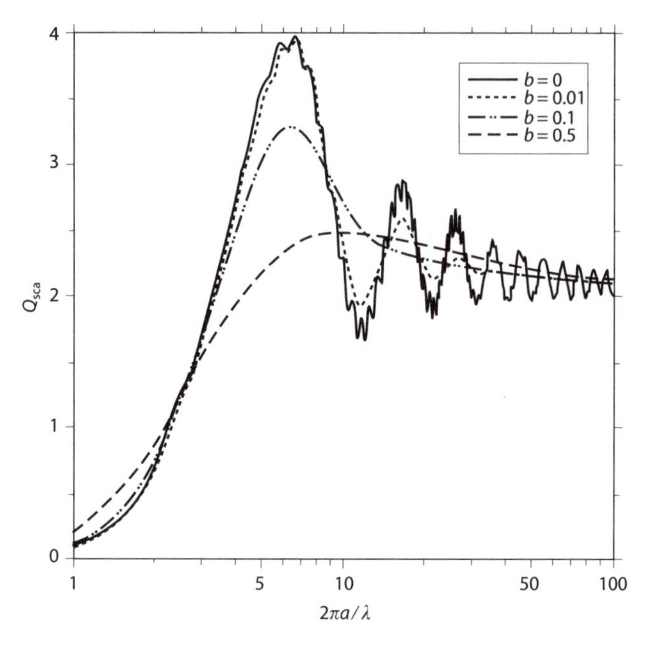

Figure 8.15 The scattering efficiency factor Q_{scat} as a function of the size parameter $x = 2\pi r/\lambda$. The refractive index $n_r = 1.33$ and complex index of refraction $n_i = 0$ are those for liquid water. Different cases of the width of the size distribution are shown where b is the effective variance (see [14]). For increasing effective variance the interference effects diminish. Adapted from [14].

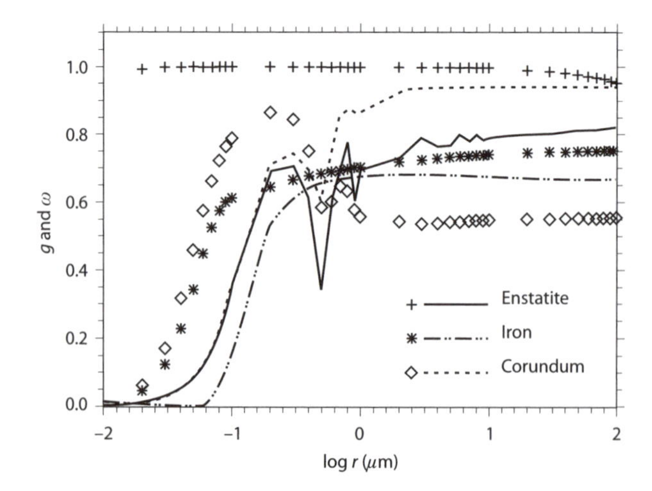

Figure 8.16 Scattering asymmetry parameter (lines) and single scattering albedo (symbols) for the three condensates described in Figure 8.17. From [17].

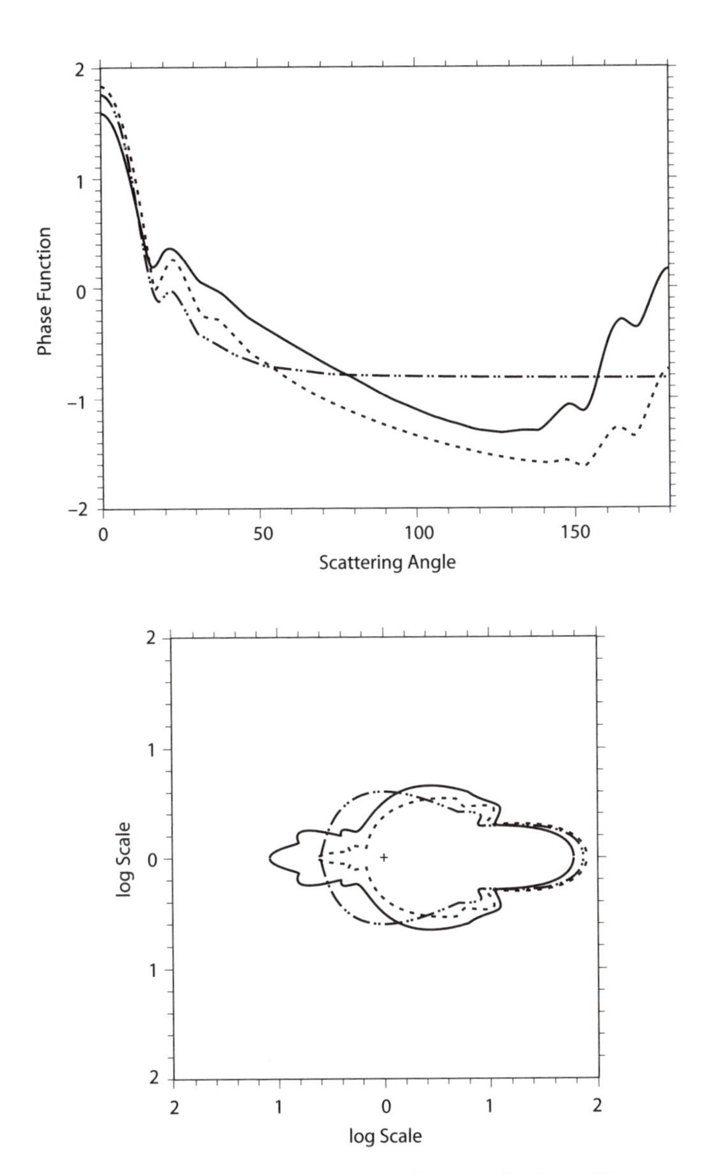

Figure 8.17 Scattering phase functions and polar diagrams for three different condensates with $\overline{r} = 1\ \mu$m, at a wavelength of 500 nm. The solid line is for $MgSiO_3$, the dot-dashed line for Fe, and the dashed line for Al_2O_3. In the polar diagram the light is incoming from the left and the condensate particle is marked by the cross. The axes on the figures are on a log scale; the units are dimensionless and only the relative numbers are important. From [17].

8.6 SUMMARY

We have delved into the details of opacity: energy level populations and absorption cross sections. Given how important opacity is in controlling the way that radiation travels through the atmosphere, we approached the topic with the goal of a conceptual understanding. We started with a review of the hydrogen atom and the Schrodinger equation and solutions. We gave the canonical description of the sources of molecular rotational and vibrational spectra, according to quantum mechanics and the classical rigid rotor and simple harmonic oscillator. Condensates (i.e., cloud or haze particles) are aggregrates of molecules and therefore lack the well-defined energy transitions that molecules have. Mie theory is the appropriate description of scattered and absorbed (and hence emitted) radiation from solid particles. We outlined the formalism used in Mie theory calculations, and a tour of some of the essential results. An understanding of the origin of opacity, our umbrella term for absorption, emission, and scattering, completes our foray into the connection of the emergent spectrum to radiation traveling through an exoplanetary atmosphere. We are now ready to move on to explore the vertical thermal structure of an atmosphere, its origin, and its relation to radiation in the atmosphere.

REFERENCES

For further reading

Further reading for quantum mechanics of molecular spectroscopy:

- French, A. P., and Taylor, E. F. 1978. *An Introduction to Quantum Physics* (New York: W. W. Norton and Co.).

- Hanel, R. A., Conrath, B. J., Jennings, D. E., and Samuelson, R. E. 2003. *Exploration of the Solar System by Infrared Remote Sensing* (2nd ed; Cambridge: Cambridge University Press).

A very detailed physics reference handbook:

- Drake, G. W. F. 2006. *Springer Handbook of Atomic, Molecular, and Optical Physics* (New York: Springer).

Further reading for scattering processes:

- van de Hulst, H. C. 1981. *Light Scattering by Small Particles* (New York: Dover).

- Bohren, C. F., and Huffmann, D. R. 1983. *Absorption and Scattering of Light by Small Particles* (New York: Wiley-Interscience).

- Hansen, J. E., and Travis, L. D., 1974. "Light Scattering in Planetary Atmospheres." Space Sci. Rev. 16, 527–610.

References for this chapter

1. R. Rutten lecture notes,
http://www.astro.uio.no/ matsc/school-07/rutten/

2. Hyperphysics molecular energy level illustration,
http://hyperphysics.phy-astr.gsu.edu/hbase/molecule/molec.html#c1

3. Hanel, R. A., Conrath, B. J., Jennings, D. E., and Samuelson, R. E. 2003. *Exploration of the Solar System by Infrared Remote Sensing* (2nd ed; Cambridge: Cambridge University Press).

4. Barrow, G. M. 1963. *The Structure of Molecules* (New York: W. A. Benjamin).

5. Drake, G. W. F. 2006. *Springer Handbook of Atomic, Molecular, and Optical Physics* (New York: Springer).

6. Sharpe, S. W., Johnson, T. J., Sams, R. L., Chu, P. M., Rhoderick, G. C., Johnson, P. A. 2004. "Gas-Phase Databases for Quantitative Infrared Spectroscopy." Appl. Spec. 58, 1452–1461.

7. Chamberlain, J. W., and Hunten D. 1978. *Theory of Planetary Atmospheres* (New York: Academic Press).

8. Thomas, G. E., and Stammes, K., 1999. *Radiative Transfer in the Atmosphere and Ocean* (Cambridge: Cambridge University Press).

9. Mihalas D. 1978. *Stellar Atmospheres* (2nd ed.; San Francisco: W. H. Freeman).

10. Liou, K. N. 2002. *An Introduction to Atmospheric Radiation* (London: Academic Press).

11. Rothman, L., et al. 1998. "The HITRAN Molecular Spectroscopic Database and HAWKS (HITRAN Atmospheric Workstation: 1996 Edition." J. Quant. Spec. Rad. Trans. 60, 665–710.

12. Bates, D. R. 1984. "Rayleigh Scattering by Air." Planet Space Sci. 32, 785–790.

13. Dalgarno, A., and Williams, D. A. 1962. "Rayleigh Scattering by Molecular Hydrogen." Astrophys. J. 1962. 136, 690–692.

14. Hansen, J. E., and Travis, L. D. 1974. "Light Scattering in Planetary Atmospheres." Space Sci. Rev. 16, 527–610.

15. Bohren, C. F., and Huffmann, D. R. 1983. *Absorption and Scattering of Light by Small Particles* (New York: Wiley-Interscience).

16. van de Hulst, H. C. 1981. *Light Scattering by Small Particles* (New York: Dover).

17. Seager, S., Whitney, B., and Sasselov D. D. 2000. "Photometric Light Curves and Polarization of Close-In Extrasolar Giant Planets." Astrophys. J. 540, 504–520.

EXERCISES

1. The Schrodinger equation. The 3D spherically symmetric Schrodinger equation can be written

$$\frac{-\hbar^2}{2m}\nabla^2\Psi(r,\theta,\phi)+U(r,\theta,\phi)\Psi(r,\theta,\phi)=E\Psi(r,\theta,\phi). \tag{8.84}$$

Show that the 3D Schrodinger equation can be expanded in polar spherical coordinates to take the form

$$-\frac{\hbar^2}{2\mu}\frac{1}{r^2\sin\theta}\left[\sin\theta\frac{\partial}{\partial r}\left(r^2\frac{\partial\Psi}{\partial r}\right)+\frac{\partial}{\partial\theta}\left(\sin\theta\frac{\partial\Psi}{\partial\theta}\right)+\frac{1}{\sin\theta}\frac{\partial^2\Psi}{\partial\phi^2}\right]$$
$$+U(r)\Psi(r,\theta,\phi)=U(r)\Psi(r,\theta,\phi). \tag{8.85}$$

2. Three gases make up 99.96% of dry air by volume: N_2 (78.1%), O_2 (20.9%), and Ar (0.93%). The major greenhouse gases are present in much smaller amounts: H_2O (less than 4%), CO_2 (385 ppm), CH_4 (1.7 ppm), and N_2O ($\sim$320 ppb). Explain why, despite their very low abundances, these greenhouse gases are such strong absorbers. Give as detailed a qualitative statement as possible.

3. Imagine you are trying to create a designer coolant that has no negative effects on the atmosphere. For example, CFCs are not only ozone-depleting gases but also greenhouse gases. In order not to contribute to warming of the Earth, which properties in a molecule would you try to avoid? Give as detailed a qualitative statement as possible.

4. Why does the molecule TiO have such strong absorption features in M stars (see Figure 3.3), when the solar abundance of Ti is so small?

5. The rotational spectrum of a diatomic molecule can be used to calculate internuclear bond lengths. Estimate the internuclear separation of CO by estimating the spacing between the rotational lines in Figure 8.7.

6. How would the molecular rotational and vibrational spectra look with the addition of different isotopes of the same molecule?

7. Show that at room temperature almost all molecules are in the ground vibrational state. Use equation [8.32]. What happens at temperatures of the hot super Earths, around 1000 K?

8. Consider a small volume of gas in complete thermodynamic equilibrium, where the intensity is Planckian: $I = J = B$. Using the detailed balance equation for an upper and lower excited state (equation [8.33]), derive the relationship between the Einstein coefficients. Why do the relationships between the Einstein coefficients still hold outside of thermal equilibrium?

9. Line broadening.

   a. Show that the line profiles for Doppler and Lorentz broadening are normalized.

   b. At which altitude in Earth's atmosphere does Doppler broadening become more significant than Lorentz broadening? Use a typical HWHM for Lorentz broadening as $\alpha = 0.07$ cm^{-1}. Use any reasonable reference for Earth's atmospheric temperature and pressure.

   c. Which broadening term dominates, if any, in the temperature range of hot Jupiter atmospheres (e.g., 1000–2000 K over the range of 10^6 to 1 Pa)?

10. A planet orbiting a distant star made news headlines around the world in spring 2007 when it was announced to be "potentially habitable." Potentially habitable means that the planet has the right surface temperature to have liquid water (and hence might support life). Yet all scientists could measure was the planetary mass and the distance from the host star.

    a. Estimate the planet's surface temperature and show that this planet is probably too hot to have surface liquid water. Use the temperature of the host star (3500 K) and the planet-star separation (0.07 AU, where 1 AU = 1.5×10^8 km). Assume the planet has the radius of Earth and the star's radius is 278,200 km. Consider that the planet has an Earth-like atmosphere.

    b. Find a scenario using nonabsorbing atmospheric aerosols that can make the surface temperature of the planet suitable for liquid water. Assume that the aerosols are spherical and with density = 0.8 or 1.2 g/cm^3, real refractive index = 1.3 or 1.5, and imaginary refractive index = 0.0. The radius of the aerosols is your choice. Compute the single scattering albedo, scattering efficiency, scattering cross section, and forward scattering asymmetry parameter for these particles at the average stellar wavelength 0.8 μm. Estimate the total mass of aerosols needed. Are nonabsorbing aerosols a plausible scenario for this planet to be "potentially habitable"? You will need a Mie theory computer code to solve this part of the problem.

11. Summary of the emission coefficient for scattering. In equation [8.1] we wrote a summary form of the absorption coefficient. Write a similar summary of the emission coefficient, including the angle dependence of scattering.

Chapter Nine

Vertical Thermal Structure of a Planetary Atmosphere

9.1 INTRODUCTION

The vertical temperature structure of the planet atmosphere is fundamental to understanding exoplanets for several different reasons. The most significant reason is that the temperature of the planetary surface tells us whether or not a planet is habitable. If the surface is too hot for covalent bonds to form complex molecules, then life cannot exist. More conventionally, people believe that liquid water is necessary at the surface for life to exist. The surface temperature and pressure define whether or not liquid water is stable. Beyond habitability, the vertical temperature pressure structure is the starting point for computing equilibrium and nonequilibrium chemistry to understand the composition of the planetary atmosphere.

Each of the solar system planet atmospheres has a qualitatively similar vertical temperature profile (Figure 9.1). This motivates us to begin with an understanding of the vertical temperature profile. Atmospheric temperatures can and do vary in the horizontal as well as in the vertical direction. This is especially true for hot exoplanets tidally locked to have permanent day and night sides. We defer a discussion of horizontal heat transport until the next chapter.

In this chapter we describe the 1D vertical pressure and temperature structure of a planet atmosphere. It is common to use pressure as a proxy for altitude (they are connected by hydrostatic equilibrium). We will use the abbrievation T-P profile. There is no simple equation to describe the vertical T-P structure, but rather many physical ingredients are at play.

9.2 EARTH'S VERTICAL ATMOSPHERIC STRUCTURE

We will begin with an overview of Earth's vertical temperature, pressure, and density structure, simply because Earth's atmosphere is the most accessible planetary atmosphere to us. Figure 9.2 shows the Earth's vertical structure typical of midlatitudes.

Earth has several different regions of the atmosphere, the lower four corresponding to altitudes of temperature reversals. These are regions of differing temperature profiles, caused by absorption of and subsequent heating by solar radiation. Solar radiation is absorbed at different altitudes, depending on the wavelength of light

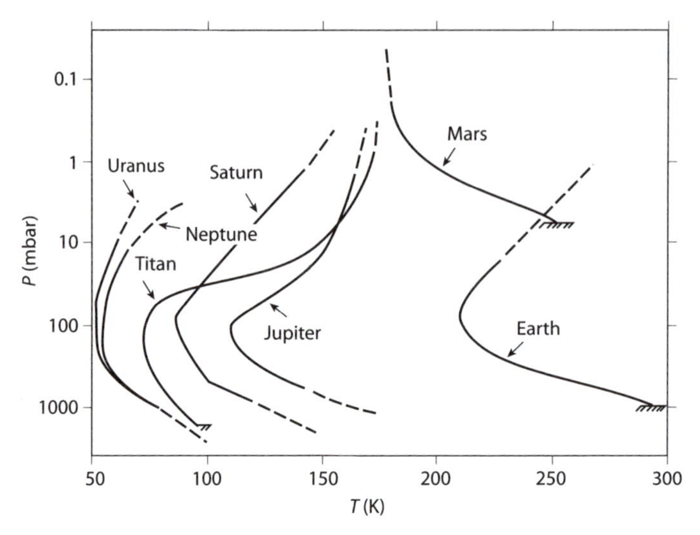

Figure 9.1 The vertical structure of solar system planet atmosphere temperatures. Notice the overall similarity in the temperature profiles. The thermal inversion, or stratosphere, on most planets comes from CH_4-induced hazes for the giant planets and from O_3 for Earth. Adapted from [1].

and the atmospheric composition. In Chapter 6 we described how radiation penetrates to different altitudes depending on the wavelength. For Earth, see Figure 9.3.

The lowest layer of Earth's atmosphere is the "troposphere." This is the region where we live and where most of what we call the weather occurs. For scale, note that Mt. Everest, at 8,856 m, reaches less than halfway through the troposphere. Most of the mass of Earth's atmosphere is in the troposphere—around 85%—because the atmospheric density decreases exponentially with increasing altitude (see the left-hand y-axis of Figure 9.2). The exponentially decreasing density is caused by hydrostatic equilibrium (Section 9.3.1) and is also familiar to Mt. Everest climbers, who require bottled oxygen to combat the decreasing amount of oxygen as they ascend. Because the troposphere contains the highest density, the troposphere is where most of the strong spectral features at visible and infrared wavelengths originate.

In the troposphere the temperature is hottest at the ground and decreases with increasing altitude. The ground is heated by solar radiation that makes it through the atmosphere unimpeded. We see from Figure 9.3 that most of the visible-wavelength solar radiation penetrates Earth's atmosphere to the surface. The ground reradiates the absorbed solar energy to heat the atmosphere. Some solar radiation is absorbed in the troposphere (by water vapor, carbon dioxide, and other gases at infrared wavelengths). The greenhouse effect also plays a role.

The stratosphere is the layer above the troposphere. There is a temperature inversion in the stratosphere. In other words, the temperature increases with increasing altitude, opposite to the troposphere. The stratopause marks the beginning of the temperature inversion. Why is there a temperature inversion in the strato-

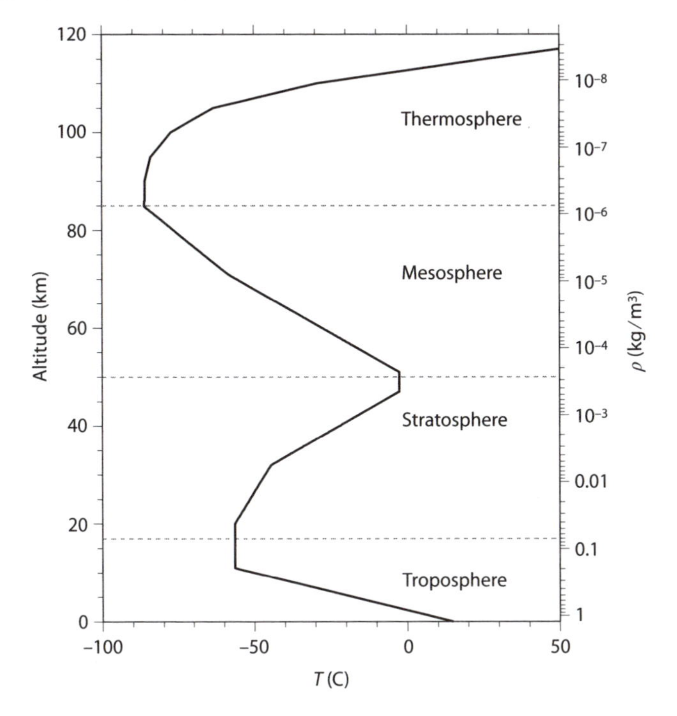

Figure 9.2 The vertical structure of Earth's atmosphere typical of midlatitudes. From the 1976 U.S. Standard Atmosphere [2]. Notice how the highest densities near the surface imply that most of Earth's mass is in the troposphere.

sphere? The stratosphere is heated from above by absorption of UV solar radiation by ozone. This UV radiation is detrimental to most biological cells, making the stratosphere a protective layer to life on Earth.

The stratosphere and the troposphere are the layers of the atmosphere that are most relevant for us. They are the regions where the spectral features occur and in turn where the surface lies and weather happens. The upper layers do have some effect on the lower levels, via photochemistry and atmospheric escape (Chapter 4).

9.3 HYDROSTATIC EQUILIBRIUM AND THE PRESSURE SCALE HEIGHT

Eighty-five percent of the mass of Earth's atmosphere is in the lower 10 km, out of an atmosphere that extends to about 100 km. Mountain climbers know this; for example, most climbers require bottled oxygen when hiking up Mt. Everest. Planetary atmospheres have an exponentially decreasing density and pressure with increasing altitude. In this section we derive hydrostatic equilibrium and the pressure scale height to show why.

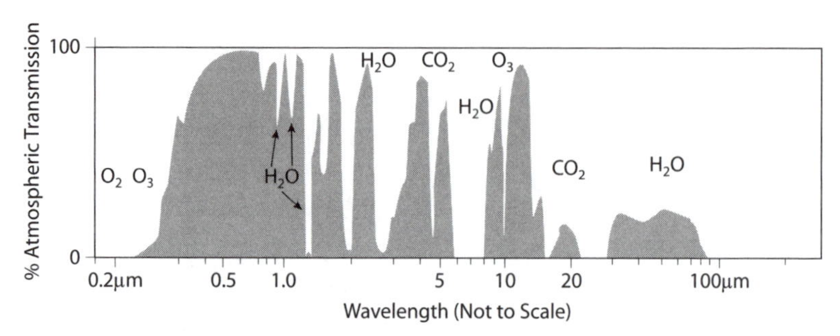

Figure 9.3 Transmission of solar radiation. Adapted from [1] and references therein.

9.3.1 Hydrostatic Equilibrium

Hydrostatic equilibrium is the balance between gravitational and pressure forces. Gravity acts to compress the atmosphere. The gas pressure balances this compression. In the case of a planet atmosphere, hydrostatic equilibrium is a balance between the outward pressure gradient force and inward gravitational force from the weight of the overlying material. The "static" in hydrostatic equilibrium refers to stationary conditions.

Hydrostatic equilibrium is required for a planet atmosphere to be stable. For example, Earth's atmosphere would collapse into a thin shell if it weren't for the outward pressure force. Without the force of gravity, the gas in a planetary atmosphere would diffuse into space, leaving an atmosphereless planet.

We now proceed with the derivation of the hydrostatic equilibrium equation. We consider a volume element of homogeneous gas where the pressure forces $F_P(z)$ in the volume element balance the gravitational forces $F_g(z)$. Here δV is the volume element, δA is the cross-sectional area, and δz is the distance along the column. We define the sign convention as positive downward toward the planetary surface.

The gravitational force on the volume element is described by

$$F_g(z) = mg = \rho g \delta V = \rho g \delta z \delta A, \tag{9.1}$$

where g is the gravitational acceleration. The pressure force

$$F_P(z) = dP\delta A \tag{9.2}$$

is related to the pressure difference dP between the pressure on the upper and lower surfaces of the volume element:

$$dP = P(z_1 + dz) - P(z_1) = P(z_1) + \frac{dP}{dz}\delta z - P(z_1) = \frac{dP}{dz}\delta z, \tag{9.3}$$

where the first term in the second equality is expanded to first order using Taylor's expansion. Equating the gravitational and pressure forces from the above three equations, we obtain

$$\frac{dP}{dz}\delta z \delta A = -g\rho \delta z \delta A. \tag{9.4}$$

Here the negative sign arises because the gravitational force is acting downward on the planetary volume element and the pressure force is acting upward, and we have set the net force equal to zero. In the limit as dz becomes infinitesimally small, we have the hydrostatic equilibrium equation,

$$\boxed{\frac{dP}{dz} = -g\rho.} \tag{9.5}$$

We have previously described a planetary atmosphere in terms of an optical depth scale rather than an altitude scale. To convert the hydrostatic equilibrium equation from an altitude scale to an optical depth scale, we use the definition of optical depth (equation [5.14])

$$\boxed{\frac{dP}{d\tau} = \frac{g\rho}{\overline{\kappa}}.} \tag{9.6}$$

Here we have used the mean optical depth scale together with the mean opacity $\overline{\kappa}$, but any reference scale is adequate as long as τ_ν and κ_ν correspond to each other at the specified frequency.

Hydrostatic equilibrium means that the atmosphere is stable; it is neither collapsing, nor expanding, nor escaping. For almost all situations in a planetary atmosphere, the above hydrostatic equilibrium equation is the form we want to use.

One assumption we have made is in omitting any vertical acceleration, whereby the net force on a volume element need not be zero. There are situations where vertical accelerations do occur, for example, hydrodynamic escape in the very upper atmosphere of an evolving planetary atmosphere. Another example is on very small scales—much smaller than we are considering in this book. Vigorous small-scale systems, such as tornadoes, thunderstorms, and convection, have nonzero vertical acceleration.

9.3.2 The Equation of State

The equation of state relates the temperature, pressure, and density of a given material. For planetary atmospheres it is adequate to use the ideal gas law

$$P = nkT = \frac{\rho kT}{\mu_{\rm m} m_{\rm H}}. \tag{9.7}$$

Here n is the number density, k is Boltzmann's constant and $\mu_{\rm m}$ is the mean molecular weight,

$$\mu_{\rm m} \equiv \frac{\overline{m}}{m_{\rm H}}, \tag{9.8}$$

where $\overline{m}$ is the number-weighted average mass of the molecules and atoms in the gas and $m_{\rm H}$ is the mass of the hydrogen atom.

9.3.3 The Pressure Scale Height

The pressure scale height H is a characteristic length scale of the planetary atmosphere. H is important because we can use it to estimate the total height and volume of the atmosphere.

As we shall see, the pressure scale height is the e-folding distance for pressure for an isothermal atmosphere with a constant mean molecular weight. The pressure scale height, in other words, is the altitude above which the pressure drops by a factor of e (e $= 2.71$). The pressure scale height is a characteristic length which is a good measure for the radial extent of a planetary atmosphere—the spectra we can observe come from a region in the atmosphere limited to several scale heights.

We derive H from the hydrostatic equilibrium equation [9.5]

$$\frac{dP}{dz} = -g\rho,$$

and the ideal gas law (equation [9.7]

$$P = nkT = \frac{\rho kT}{\mu_{\rm m} m_{\rm H}}.$$

Combining these equations gives

$$\frac{dP}{P} = -\frac{\mu_{\rm m} m_{\rm H} g}{kT} dz \tag{9.9}$$

with a solution

$$P = P_0 {\rm e}^{-z/H}, \tag{9.10}$$

where we have defined the pressure scale height to be

$$\boxed{H \equiv \frac{kT}{\mu_{\rm m} m_{\rm H} g}.} \tag{9.11}$$

Here again $m_{\rm H}$ is the mass of the hydrogen atom and $\mu_{\rm m}$ is the mean molecular weight. We emphasize that the pressure scale height we have derived is for constant T and constant $\mu_{\rm m}$, and we have neglected the variation of g with altitude.

The scale height H for solar system planets is on the order of 5 to 20 km, and is given in Table 3.1. For hot Jupiter exoplanets, the scale height can be several hundred kilometers. We may estimate the total atmosphere relevant for spectral lines as approximately 5 scale heights.

With the pressure scale height we can also get a handle on how the pressure and density vary as a function of altitude. Even though we have assumed an isothermal atmosphere, equation [9.11] shows us that the pressure (and density via the ideal gas law) varies exponentially with altitude. The temperature for a typical planetary atmosphere varies by only a factor of a few (see Figure 9.1).

9.4 SURFACE TEMPERATURE FOR A SIMPLIFIED ATMOSPHERE

We begin our investigation into the vertical thermal structure of a planetary atmosphere by estimating the difference between the atmosphere and surface temperatures in a very simple model. We will use a previous derivation of the equilibrium temperature $T_{\rm eq}$ in Chapter 3. In Chapter 3 and Figure 3.1 we used the concept of energy balance to equate the total flux from a star incident on a planet (equation [3.3]),

$$(1 - A_{\rm B})\mathcal{F}_{{\rm S},*} \left(\frac{R_*}{a}\right)^2 \pi R_{\rm p}^2, \tag{9.12}$$

with the total flux emerging from a planet,

$$4\pi R_{\rm p}^2 \mathcal{F}_{\rm S,p}, \tag{9.13}$$

to find

$$\sigma_{\rm R} T_{\rm eq}^4 = \frac{1}{4}\sigma_{\rm R} T_{\rm eff,*}^4 \left(\frac{R_*}{a}\right)^2 (1 - A_{\rm B}) \tag{9.14}$$

and the equilibrium temperature (equation [3.3])

$$T_{\rm eq} = T_{\rm eff,*}\left(\frac{R_*}{2a}\right)^{1/2} (1 - A_{\rm B})^{1/4}. \tag{9.15}$$

Here we have assumed that the absorbed stellar radiation is circulated evenly around the planet. We have also used the Stefan-Boltzmann Law (equation [3.6]), equating flux with temperature $\mathcal{F} = \sigma_{\rm R} T^4$ where $\sigma_{\rm R}$ is the radiation constant and T is temperature.

Recall that $T_{\rm eq}$ is a theoretical number that is the temperature attained by an isothermal planet after it has reached complete equilibrium with the radiation from its parent star. In the context of an idealized planetary atmosphere, the equilibrium temperature is essentially the temperature at the layer where most of the radiation is emitted. For this subsection we will refer to the equilibrium temperature as the emission temperature, $T_{\rm eq} = T_{\rm e}$. We also denote the stellar effective temperature $T_{\rm eff,*} = T_*$.

We now move to describe a simple greenhouse atmosphere (Figure 9.4), still considering uniform redistribution of absorbed stellar radiation, and following [3]. In this atmosphere there are two layers. The bottom layer is the surface. We assume that all of the radiation from the star (in the form of stellar flux) reaches the surface, that is, there is no scattering. This is a reasonable approximation for stars with most of their energy output at visible wavelengths and for a planet atmosphere composed of molecules that primarily absorb and emit at infrared wavelengths. We assume the radiation that reaches the ground is absorbed, heats the surface, and is reemitted at longer wavelengths characteristic of the atmospheric temperature (see the discussion of black body radiation in Section 2.8 and Figure 2.7).

The reprocessed flux (in the amount of the absorbed incident stellar flux) then travels upward and is absorbed by the second layer. We will call this second layer the atmosphere layer. In this picture, the atmospheric layer absorbs at infrared wavelengths; we have already assumed the planetary atmosphere is composed of molecules that primarily emit at IR wavelengths.

We assign a temperature $T_{\rm a}$ to the atmosphere layer and $T_{\rm s}$ to the surface layer. Our goal is to estimate the surface temperature in terms of the planet's equilibrium temperature ($T_{\rm e}$) and in terms of the atmosphere layer temperature. Let us focus first on the atmospheric layer. From energy balance, the radiation emerging from the top of the atmosphere layer must be equivalent to the absorbed stellar radiation,

$$\sigma_{\rm R} T_{\rm a}^4 = \sigma_{\rm R} T_*^4 \left(\frac{R_*}{2a}\right)^2 (1 - A_{\rm B}). \tag{9.16}$$

But the definition of equilibrium temperature $T_{\rm e}$ comes from the same energy balance requirement (equation [9.15]). We therefore equate the atmospheric and emission temperatures,

$$\sigma_{\rm R} T_{\rm a}^4 = \sigma_{\rm R} T_{\rm e}^4. \tag{9.17}$$

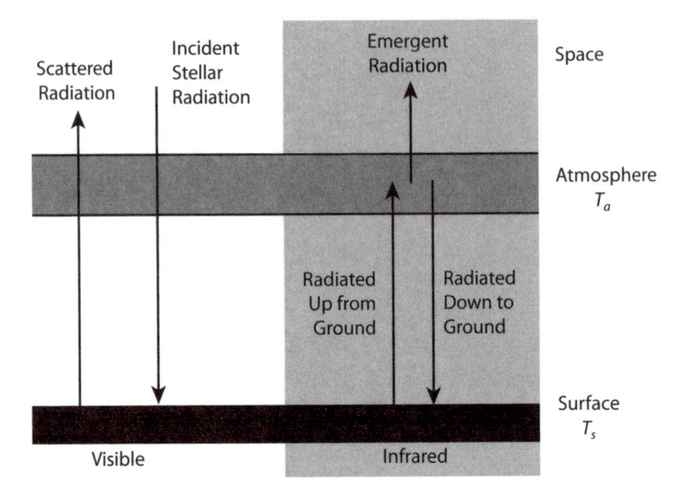

Figure 9.4 The simple greenhouse. Adapted from [3].

The atmosphere layer is the layer emitting to space, so it makes sense that the atmosphere and emission temperatures are equivalent.

At the surface, energy balance again means that all of the incoming stellar radiation absorbed must be subsequently reradiated. There are two contributions to the absorbed radiation at the surface: the radiation incoming from the star and the downward radiation from the atmosphere layer,

$$\sigma_{\mathrm{R}} T_{\mathrm{s}}^4 = \sigma_{\mathrm{R}} T_*^4 \left(\frac{R_*}{2a}\right)^2 (1 - A_{\mathrm{B}}) + \sigma_{\mathrm{R}} T_{\mathrm{a}}^4. \tag{9.18}$$

We have previously shown above that the first term on the right-hand side is equivalent to $\sigma_{\mathrm{R}} T_{\mathrm{e}}^4$ and the second term on the right-hand side is also equivalent to $\sigma_{\mathrm{R}} T_{\mathrm{e}}^4$, giving

$$\boxed{T_{\mathrm{s}} = 2^{1/4} T_{\mathrm{e}}.} \tag{9.19}$$

We see that the surface temperature is about 1.19 times the atmospheric temperature. In this simplified greenhouse model, the atmosphere layer contributes to heating the planet surface. This simple greenhouse model overestimates Earth's surface temperature but significantly underestimates Venus's surface temperature. For Earth the model gives $T_{\mathrm{s}} \sim 303$ K, based on $T_{\mathrm{e}} \sim 255$ K, compared to the average surface temperature $T_{\mathrm{s}} \sim 280$ K. For Venus, the model surface temperature is $T_{\mathrm{s}} \sim 274$ K based on $T_{\mathrm{e}} = 230$ K, compared to the measured $T_{\mathrm{s}} = 730$ K.

In the simple greenhouse model we have assumed that the absorbed stellar radiation is trapped by the atmospheric layer. In a slightly more realistic model we consider a "leaky" greenhouse [3]. In the leaky greenhouse (Figure 9.5) the single atmosphere layer is optically thin. Again, we make the simplifying assumption that all of the radiation from the star not scattered back to space reaches the planet surface. We will write down two equations to solve for T_{s} in terms of T_{e}.

By energy balance the radiation leaving the planet must be equal to the total amount of radiation absorbed by the planet. There are two contributions to the

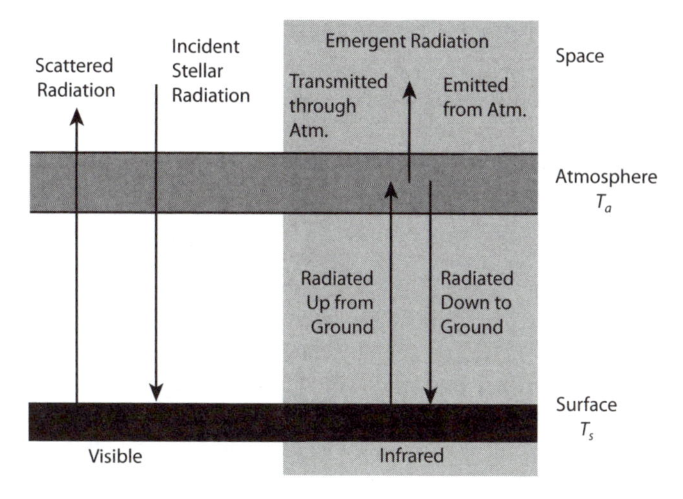

Figure 9.5 The leaky greenhouse. Adapted from [3].

radiation leaving the planet. The first contribution is from the atmosphere layer and the second contribution is the surface radiation that leaks through the atmosphere. We therefore have

$$\sigma_{\mathrm{R}} T_{\mathrm{e}}^4 = \sigma_{\mathrm{R}} T_{\mathrm{a}}^4 + \sigma_{\mathrm{R}} T_{\mathrm{s}}^4 (1 - \alpha), \tag{9.20}$$

where $\alpha \in (0, 1)$ is the fraction of radiation that is absorbed by the atmosphere layer. At the planet surface, we again have

$$\sigma_{\mathrm{R}} T_{\mathrm{s}}^4 = \sigma_{\mathrm{R}} T_{\mathrm{a}}^4 + \sigma_{\mathrm{R}} T_{\mathrm{e}}^4. \tag{9.21}$$

Solution of the above two equations yields

$$\boxed{T_{\mathrm{s}} = \left(\frac{2}{2 - \alpha} \right)^{1/4} T_{\mathrm{e}}.} \tag{9.22}$$

The greenhouse warming effect is reduced by the paritally transparent atmosphere layer. We can see that if $\alpha = 1$ all radiation is trapped by the atmosphere layer, in agreement with the simple greenhouse model (also showing that the simple greenhouse model is a special case of the more general leaky atmosphere model). In the limit $\alpha \to 0$, we recover $T_{\mathrm{s}} = T_{\mathrm{e}}$, the case where the atmosphere layer is transparent to radiation (or there is no atmosphere layer). The surface layer is always warmer than the atmosphere layer in the leaky greenhouse model. We could further show, using Kirchhoff's Law, that $T_{\mathrm{s}} \geq T_{\mathrm{e}} \geq T_{\mathrm{a}}$.

In the above simple greenhouse and leaky greenhouse models we have effectively explored a radiative equilibrium model for a one-layer atmosphere in the case of no scattering. If we were to extend the simple models to many layers, with the appropriate value for α in each layer, considering all wavelengths, molecules, and clouds, we would approach a multilayer radiative equilibrium model.

9.5 CONVECTION VERSUS RADIATION

In a planetary atmosphere energy can be transported by radiation, convection, or thermal conduction. The competing heat transport methods are radiation and convection. In the relatively dense lower atmospheres that we are considering, thermal conduction is not important. What determines whether or not convection is taking place in a planetary atmosphere? Energy will be transported by the most efficient method—that is, via the "path of least resistance." If the opacity (the absorption or scattering of photons) is low, then the energy is most likely to be transported by photons. If the opacity is high then the resistance to the flow of photons is high, and energy is more likely to be carried by convection.

To quantify which energy transport mechanism dominates, we turn to a discussion of temperature gradients (dT/dz). We begin with a qualitative discussion on how the temperature gradient determines whether or not convection will occur. If convection will occur we call the atmosphere unstable against convection. We call a situation stable if, after a disturbance, the system will return to its original state.

Consider an air parcel that is slightly unstable, rises a slight distance, and expands adiabatically to the ambient pressure. Adiabatic expansion means that there is no heat exchange with its surroundings; the air parcel will have the same pressure as the surrounding atmosphere, but its own (possibly different) temperature and density. If the air parcel is colder and therefore more dense than its surroundings it will sink, and convection will not occur. The atmosphere is said to be stable against convection. In contrast, the atmosphere is said to be unstable against convection if, after rising a slight distance and expanding, the air parcel continues to rise. This happens if the air parcel is hotter and therefore less dense than its surroundings, so that the buoyancy force will cause the air parcel to continue to rise.

The criterion for convection, then, is related to two temperature gradients: (1) the adiabatic temperature gradient (the temperature gradient followed by the rising air parcel) and (2) the surrounding atmospheric temperature gradient. If the adiabatic temperature gradient is shallower than the atmospheric temperature gradient, the atmosphere is unstable and convection occurs. This criterion for convection is then a criterion for buoyancy,

$$\left(\frac{dT}{dz}\right)_{\rm ad} > \left(\frac{dT}{dz}\right)_{\rm atm}. \tag{9.23}$$

Convective stability and instability in terms of the atmospheric and adiabatic temperature gradients are described in Figure 9.6.

In order to understand under what conditions a temperature gradient is small or large we now turn to a quantitative discussion of the adiabatic and radiative temperature gradients. The adiabatic temperature gradient is (derived in Section 9.7.2)

$$\frac{dT}{dz} = -\frac{g}{c_p}, \tag{9.24}$$

where g is the surface gravity and c_p is the specific heat capacity at constant pressure in units of J kg^{-1} s^{-1}.

Convection will set in when the adiabatic gradient becomes small, or when the radiative temperature gradient becomes large. The adiabatic temperature gradient

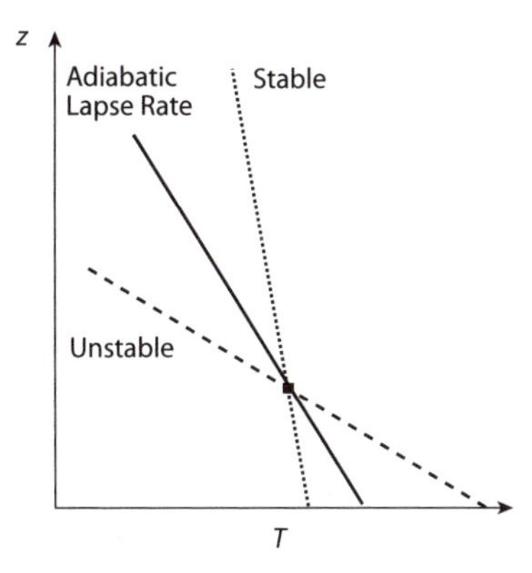

Figure 9.6 Illustration of temperature profiles stable or unstable against convection.

(equation [9.24]) becomes small when the heat capacity becomes large. The specific heat capacity is the amount of energy that must be transferred to the gas per unit mass at constant pressure to increase the temperature by 1 K. In a situation where a large amount of energy is needed to heat the gas by 1 K, the adiabatic temperature gradient will become very small. A typical example in stars is the layer where an abundant element such as hydrogen starts to ionize. The temperature increase goes into energy to remove the electrons from the atoms, rather than to increase the kinetic energy in the gas; the heat capacity becomes large because a lot of energy is needed to heat the gas. A similar situation also occurs in the hot interiors of giant planets like Jupiter, where at $T > 4000$ K hydrogen starts to ionize.

The radiative equilibrium temperature gradient in the limit of the diffusion approximation is (Section 6.4.4)

$$\left(\frac{dT}{dz}\right)_{\rm rad} = -\frac{3}{16} F_{\rm int} \frac{\overline{\kappa}}{\sigma_{\rm R} T^3}. \tag{9.25}$$

While the above equation is valid only deep in a planetary atmosphere where the diffusion approximation holds, the equation serves to illustrate criteria where the radiative temperature gradient becomes large. The radiative temperature gradient becomes large when either the radiative flux F or the mean opacity κ becomes large. In planet interiors, the fluxes may be large for giant planets, driving convective interiors. In the atmospheres, the fluxes may not be large enough to force convection.

The opacity, on the other hand, is a driving factor for convection. The opacity becomes large where the number density of absorbing particles becomes large. Hence, deep in a planetary atmosphere where high opacities and optically thick conditions prevail, convection almost always sets in. In the upper part of the planetary atmosphere, transport of energy by radiation usually dominates over convection. In addition, the opacity can change from large to small (or vice versa) due to changes in temperature and pressure versus altitude in a planetary atmosphere—these change the atomic or molecular makeup of the atmosphere. Jupiter, for example, follows the above description of a radiative zone in the upper atmosphere and a convection zone in the lower atmosphere. But, beneath the convection zone, Jupiter may have another radiative zone, because of a minimum in the mean opacities at high temperatures (1300–1800 K [4]). Hot Jupiter exoplanets are different examples of the interplay between radiative and convective regions. In contrast to Jupiter, the radiative zones in hot Jupiter atmospheres are expected to continue deep into the atmosphere on their permanent day sides because the strong heating from the parent star causes the temperature gradient to be close to isothermal or even inverted.

Planets like Earth with solid surfaces beneath relatively thin atmospheres are always expected to have convection zones just above the solid surface. The reason is that most of the visible-wavelength incident stellar energy is absorbed at the planet surface, while only some is absorbed in the planet atmosphere. This uneven energy absorption by the atmosphere and surface makes the surface hotter than the overlying atmosphere layers heated only by absorption of radiation. A significant temperature discontinuity will arise between the atmosphere and the planetary surface. This temperature discontinuity will drive convection. On Earth convection is occurring in most of the troposphere, but may be limited to a thinner layer on exoplanets with different atmospheric conditions. According to the criterion for convection (equation [9.23]) and the radiative equilibrium temperature gradient (equation [9.25]), we can think of a very large radiative flux being emitted from the ground layer. This causes the radiative temperature gradient to be large enough that convection sets in.

Atmosphere layers with temperature inversions are very stable against convection. Almost all of the solar system planets have temperature inversions in the upper atmospheres, where the temperature rises with increasing altitude from the planet surface. Recall that these temperature inversions are typically caused by absorption of UV radiation. The temperature inversion layers are stable against convection because the restoring force to a perturbed, lifted air parcel is very strong. Recall that as an air parcel rises, it expands and cools. Being cooler—and hence denser—than the surrouding temperature-inverted atmosphere (which increases with rising height), the air parcel will sink again.

In the next two sections we will describe the T-P profile in the atmosphere from radiative equilibrium and from convective equilibrium.

9.6 THE RADIATIVE EQUILIBRIUM TEMPERATURE PROFILE

We embark on a description of the 1D vertical temperature profile in the case where energy is transported only by radiation. We will show that, if we know the total amount of energy passing through the planetary atmosphere, we can derive the temperature profile. We begin by outlining the general case where there is no analytic solution.

9.6.1 Radiative Equilibrium

In a planetary atmosphere, no energy is created or destroyed. Therefore, the amount of radiation passing into a given volume of the atmosphere must equal the amount of radiation leaving that volume. This is called radiative equilibrium. We can also express radiative equilibrium as a flux constancy,

$$\frac{dF(\tau)}{d\tau} = 0. \tag{9.26}$$

Here $F(\tau)$ is the total radiative flux (equation [5.37]) integrated over all frequencies, and τ is the optical depth scale (Section 5.3). The radiative equilibrium temperature profile is the temperature profile that satisfies the radiative equilibrium constraint.

To continue with an expression for radiative equilibrium we consider the total amount of radiation absorbed in a given atmospheric volume:

$$\int_0^\infty \int_\Omega \kappa(\tau_\nu, \nu) I(\tau_\nu, \mu, \nu) d\Omega d\nu. \tag{9.27}$$

Here "total" means integrated over all angles and frequencies. Because absorption is isotropic we may integrate over the solid angle Ω to find

$$\int_0^\infty \kappa(\tau_\nu, \nu) J(\tau_\nu, \nu) d\nu. \tag{9.28}$$

The total amount of radiation emitted in a given atmospheric volume is

$$\int_0^\infty \int_\Omega \varepsilon(\tau_\nu, \mu, \nu) d\Omega d\nu. \tag{9.29}$$

If we make the assumption of isotropic emission, and consider the definition of the source function (Section 5.5 and equation [5.24]), we have for the total emission

$$\int_0^\infty \kappa(\tau_\nu, \nu) S(\tau_\nu, \nu) d\nu. \tag{9.30}$$

We may now equate the total amount of radiation absorbed (equation [9.28]) and emitted (equation [9.30]) in a given volume to find

$$\boxed{\frac{dF_{\mathrm{R}}(\tau)}{d\tau} = \int_0^\infty \kappa(\tau_\nu, \nu) \left[J(\tau_\nu, \nu) - S(\tau_\nu, \nu)\right] d\nu = 0.} \tag{9.31}$$

We now have a mathematical expression for radiative equilibrium: radiative absorption is balanced by radiative emission, in a given layer. It is important to realize that radiative emission and absorption in a given layer do not have to balance *at a given frequency*.

What is the constant flux being driven through the atmosphere? By radiative energy balance, the flux in each layer is the same as emitted at the top of the atmosphere, namely,

$$F(\tau) = \sigma_{\mathrm{R}} T_{\mathrm{int}}^4 + \mu_0 I_0. \tag{9.32}$$

The first term $\sigma_{\mathrm{R}} T_{\mathrm{int}}^4$ is the flux coming from the planetary interior that passes through the planet atmosphere. Giant planets have a source of interior energy from the gradual loss of residual gravitational potential energy from the planet's formation. Indeed, Jupiter has an internal luminosity over twice as high as its luminosity from reradiated absorbed stellar energy. Earth has an interior energy source partly from residual gravitational potential energy but mostly from decay of radioactive isotopes (of uranium, thorium, and potassium). For many planets, including Earth and the hot exoplanets in short-period orbits, the second term, the flux from the star, overwhelms the interior flux. (For a derivation of the term $\mu_0 I_0$ see Section 6.4.3.1 and equation [6.37].)

The equation of radiative equilibrium is coupled to the radiation field, through the angle-averaged quantity $J(\tau_\nu, \nu)$. The radiative equilibrium equation and the radiative transfer equation [5.39] must therefore be solved simultaneously. We emphasize that this radiative equilibrium equation shows how the opacity as a function of frequency plays into determining the temperature profile. A temperature inversion could arise naturally by solving the radiative transfer and radiative equilibrium equations together.

There is no analytic solution of the radiative equilibrium temperature profile in the general case. We now proceed with a gray atmosphere, where analytic solutions to the temperature profile are possible.

9.6.2 Gray Atmosphere Heated from Below

We will derive a temperature profile using the condition of radiative equilibrium (equation [9.31]) under the assumption of LTE and in an atmosphere with no scattering. In order to find an analytical solution, we must make a further simplification: that of a gray atmosphere. A gray atmosphere is one where the radiation quantities (i.e., intensity, absorption, and emission coefficients) do not vary with frequency.

We first consider an atmosphere heated from below. By this, we mean an atmosphere for which the source of radiation is from the interior or ground only. One example is a giant planet atmosphere, dominated by interior flux and heated as the interior flux travels through the atmosphere out to space. Another example is a planet with a thin atmosphere such as Mars (or even, approximately, Earth) which is completely transparent at visible wavelengths. The bulk of the stellar energy is at visible wavelength and reaches the planet surface. There, the stellar radiation is absorbed by and heats the surface and is reemitted at longer wavelengths related to the characteristic temperature of the planet. Terrestrial planet atmospheres are typically not transparent at IR wavelengths where molecules can absorb and reemit radiation, heating the atmosphere in the process. The infrared radiation coming from below therefore heats the atmosphere as it travels out through the atmosphere to space. We call the interior temperature $T_{\rm int}$. For the giant planet example above, $T_{\rm int} = T_{\rm eff}$. For the thin atmosphere transparent to visible radiation, $T_{\rm int}$ is the surface temperature $T_{\rm s}$.

For a gray atmosphere we use the total mean radiation field $J(\tau) = \int_0^\infty J(\tau_\nu, \nu) d\nu$, $I(\tau) = \int_0^\infty I(\tau_\nu, \nu) d\nu$, and $S(\tau) = \int_0^\infty S(\tau_\nu, \nu) d\nu$. In the gray case, the radiative equilibrium equation [9.31] becomes

$$J(\tau) = S(\tau). \tag{9.33}$$

With the assumption of LTE and the frequency-integrated black body flux, we have

$$\boxed{J(\tau) = S(\tau) = B(T(\tau)) = \frac{\sigma_R T^4}{\pi}.} \tag{9.34}$$

We now have enough background information to derive a temperature profile. Our goal is to show that the temperature profile in a gray, LTE, radiative equilbrium atmosphere with no scattering is

$$T^4 \approx \frac{3}{4} T_{\text{int}}^4 \left[\tau + 2/3\right]. \tag{9.35}$$

Using the definition of optical depth (equation [5.13]) the temperature versus optical depth can subsequently be converted to a temperature versus altitude.

We follow [5] and start with the radiative transfer equation [5.39]

$$\mu \frac{dI(\tau, \mu, \nu)}{d\tau} = I(\tau, \mu, \nu) - S(\tau, \nu).$$

Integrating over all frequencies, we have

$$\mu \frac{dI(\tau, \mu)}{d\tau} = I(\tau, \mu) - S(\tau). \tag{9.36}$$

Given $J(\tau) = S(\tau)$ for a gray radiative equilibrium atmosphere (equation [9.34]), the gray radiative transfer equation is

$$\mu \frac{dI(\tau, \mu)}{d\tau} = I(\tau, \mu) - J(\tau). \tag{9.37}$$

We now proceed to multiply the radiative equilibrium version of the gray radiative transfer equation by successive orders of μ and integrate over solid angle (the so-called moments of the radiative transfer equation). For the zeroth moment we integrate the above equation [9.37] over the range $-1 < \mu < 1$ and $0 \leq \phi < 2\pi$ to find

$$\frac{1}{4\pi} \frac{dF(\tau)}{d\tau} = J(\tau) - J(\tau) = 0. \tag{9.38}$$

This result is the flux constancy that we had already assumed at the beginning of this subsection. We will now refer to $F(\tau)$ as a constant F. The first moment, again integrating over the range $-1 < \mu < 1$ and $0 < \phi < 2\pi$, is

$$\frac{dK(\tau)}{d\tau} = \frac{1}{4\pi} F \tag{9.39}$$

with a solution using $dF(\tau)/d\tau = 0$,

$$K(\tau) = \frac{1}{4\pi} F\tau + c, \tag{9.40}$$

where c is an integration constant we wish to find. We can relate $K(\tau)$ to $J(\tau)$ by the Eddington approximation $K(\tau) = \frac{1}{3} J(\tau)$ (equation [6.34]) to find

$$J(\tau) = \frac{3}{4\pi} F\tau + 3c. \tag{9.41}$$

We use the radiative equilibrium LTE relationship in equation [9.34] for $J(\tau) = \sigma_R T^4/\pi$ and the constant flux that is passing through each layer of atmosphere $F = \sigma_R T_{\rm int}^4$ to conclude that

$$\boxed{T^4 \approx \frac{3}{4} T_{\rm int}^4 \left[\tau + 2/3\right].} \tag{9.42}$$

We leave the evaluation of $3c = \sigma T_{\rm int}^4/2\pi$ as an exercise, with a general outline here. To find c we calculate the flux emerging at the top of the atmosphere using a frequency-integrated version of the emergent flux (equation [5.46]),

$$F(0) = 2\pi \int_0^1 \int_0^\infty S(\tau,\mu) e^{-\tau/\mu} d\tau d\mu. \tag{9.43}$$

Using the condition for radiative equilibrium $S(\tau) = J(\tau)$, our solution for $J(\tau)$ in equation [9.41], and the upper boundary condition that the emergent flux at the top of the atmosphere is $\sigma_R T_{\rm int}^4$, we have to solve the equation

$$F(0) = 2\pi \int_0^1 \int_0^\infty \left[\frac{3}{4\pi} F\tau + 3c\right] e^{-\tau/\mu} d\tau d\mu \equiv \sigma_R T_{\rm int}^4, \tag{9.44}$$

in order to derive c. We reiterate that $F(\tau)$ is a constant denoted here by F and we use $F(0)$ for the same constant flux that is emerging at the top of the planet atmosphere.

What can the gray atmosphere radiative equilibrium temperature profile tell us about planetary atmospheres? First, the derived temperature profile gives us an approximate prescription for a temperature profile in a planetary atmosphere, given that we have the appropriate atmospheric composition and opacities and their dependence on temperature and pressure (Figure 9.7).

Second, the gray atmosphere radiative equilibrium temperature at the top of the atmosphere (equation [9.42] at $\tau = 0$) matches our estimate for a simple greenhouse atmosphere (equation [9.19]), $T_e \sim (1/2)^{1/4} T_{\rm int}$, where here $T_{\rm int}$ is the surface temperature.

Third, taking the derivative of our temperature profile (equation [9.42]) with respect to τ, we find a temperature gradient

$$\frac{dT}{dz} \approx -\frac{3}{16}\kappa(z)\frac{T_{\rm int}^4}{T^3}, \tag{9.45}$$

recovering the diffusion approximation temperature gradient if we associate the optical depth scale and the $\kappa(z)$ with the Rosseland mean. This temperature gradient, again, shows that either a high net radiative flux (in the form of $F = \sigma_R T_{\rm int}^4$) or a high κ generates a large temperature gradient.

9.6.3 Gray Atmosphere Heated from Above and Below

We have above just computed a temperature profile for a gray, LTE, radiative equilibrium atmosphere heated from below. Such an atmosphere could represent a giant planet with an interior energy source heating the atmosphere. Alternatively, the heating from below scenario could be a rocky planet with a surface, where all of

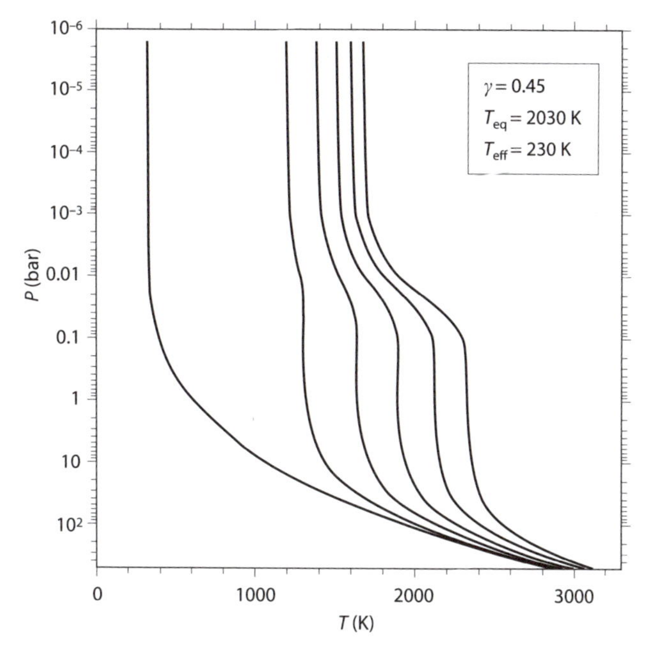

Figure 9.7 Temperature-pressure profiles computed using the gray atmosphere approximation in equation [9.62] with a scaling factor to convert from τ to P. The curves are for different incoming μ_0; from left to right for $\mu_0 = \cos\theta_0 = 0.0, 0.2, 0.4, 0.6, 0.8, 1.0$. Adapted from [5].

the incident radiation travels down through a transparent atmosphere, is absorbed by the ground, and is reradiated upward.

We now turn to the case of an atmosphere heated from above and below, where the stellar radiation plays a role throughout the atmosphere, following [5]. In order to arrive at an analytic equation for a temperature profile, we still work under the same simplifying assumptions, that of a gray atmosphere in LTE and in radiative equilibrium with no scattering. Recall that the gray atmosphere assumption means we use frequency-integrated terms of radiation quantities.

To consider heating from stellar radiation, we must consider the basic point common to all planetary atmospheres: the star is always hotter than the planet by definition. This implies that stellar radiation is absorbed at visible wavelengths and reradiated at infrared wavelengths. Molecular absorption typically dominates at IR wavelengths.

The approach to finding a temperature profile in an atmosphere heated both from above and below therefore involves two different intensities (and other quantities of radiation). One intensity is at visible wavelengths, $I_{\rm vis}$, where we assume radiation is only absorbed and not emitted. The second intensity is at infrared wavelengths $I_{\rm ir}$, where we assume that all of the absorbed energy is reradiated. The total intensity is $I_{\rm vis} + I_{\rm ir}$, and similar addition rules apply for other radiation quantities.

The motivation of the visible versus infrared separation of radiation is that the altitudes at which radiation is absorbed and emitted are different. Earth is a good example: our atmosphere, in the absence of clouds, is transparent at visible wavelengths. We know this because we can see the stars on a cloudless night. Most of the radiation at visible wavelengths is absorbed by the ground. At most IR wavelengths (with the exception of a few narrow windows), the night sky is opaque.

We will therefore assume that there is no emission at visible wavelengths, in other words $\varepsilon(\tau) = 0$, so that

$$S_{\mathrm{vis}}(\tau) \equiv \frac{\varepsilon(\tau)}{\kappa(\tau)} = 0. \tag{9.46}$$

Our radiative equilibrium constraint is still the equivalency of absorbed and emitted radiation in each atmospheric layer, but now takes on the form

$$\kappa_{\mathrm{vis}}(\tau) J_{\mathrm{vis}}(\tau) + \kappa_{\mathrm{ir}}(\tau) J_{\mathrm{ir}}(\tau) = \kappa_{\mathrm{ir}}(\tau) S_{\mathrm{ir}}(\tau), \tag{9.47}$$

which can be rewritten as

$$S_{\mathrm{ir}}(\tau) = J_{\mathrm{ir}}(\tau) + \frac{\kappa_{\mathrm{vis}}(\tau)}{\kappa_{\mathrm{ir}}(\tau)} J_{\mathrm{vis}}(\tau) = J_{\mathrm{ir}}(\tau) + \gamma J_{\mathrm{vis}}(\tau). \tag{9.48}$$

Here we have defined

$$\gamma \equiv \kappa_{\mathrm{vis}}/\kappa_{\mathrm{ir}} \tag{9.49}$$

as the ratio of the visible and infrared mean absorption coefficients.

The two radiative transfer equations for the visible and infrared beams of intenesity are

$$\mu \frac{dI_{\mathrm{vis}}(\tau, \mu)}{d\tau_{\mathrm{vis}}} = I_{\mathrm{vis}}(\tau, \mu), \tag{9.50}$$

$$\mu \frac{dI_{\mathrm{ir}}(\tau, \mu)}{d\tau_{\mathrm{ir}}} = I_{\mathrm{ir}}(\tau, \mu) - S_{\mathrm{ir}}(\tau, \mu) = I_{\mathrm{ir}}(\tau, \mu) - J_{\mathrm{ir}}(\tau) - \gamma J_{\mathrm{vis}}(\tau). \tag{9.51}$$

The zeroth-order moment equations are

$$\frac{1}{4\pi} \frac{dF_{\mathrm{vis}}(\tau)}{d\tau_{\mathrm{vis}}} = J_{\mathrm{vis}}(\tau), \tag{9.52}$$

$$\frac{1}{4\pi} \frac{dF_{\mathrm{ir}}(\tau)}{d\tau_{\mathrm{ir}}} = -\gamma J_{\mathrm{vis}}(\tau). \tag{9.53}$$

Adding the above two equations together, considering that $d\tau_{\mathrm{vis}} = \gamma d\tau_{\mathrm{ir}}$ we again see in the expression of radiative equilibrium that the total flux derivative is zero, meaning that a constant net flux passes through the atmosphere.

The first-order moment equations of the infrared radiative transfer equation are

$$\frac{dJ_{\mathrm{ir}}(\tau)}{d\tau_{\mathrm{ir}}} = \frac{3}{4\pi} F_{\mathrm{ir}}(\tau), \tag{9.54}$$

where we have used the Eddington approximation $J = 3K$ (equation [6.34]).

As before, for the gray atmosphere heated from below, we find the temperature profile by starting with an expression for $J_{\mathrm{ir}}(\tau)$. The complication compared to the case of the atmosphere heated from below is that the infrared equations also depend on the visible wavelength radiation quantities. Conceptually, this is because, although in each layer the total absorbed and emitted radiation is constant, the radiation is absorbed at visible and infrared wavelenghts, and in this framework only emitted in IR wavelengths. Explicitly, the equation for $J_{\rm ir}(\tau)$ (equation [9.54]) depends on $F_{\rm ir}(\tau)$, which itself depends on $J_{\rm vis}(\tau)$,

To find $J_{\rm vis}(\tau)$, recall from Section 6.4.3.1 that the incoming intensity from one direction μ_0, ϕ_0 is the attenuation of the incident intensity,

$$I_{\rm vis}(\tau) = I(0,\mu)e^{\gamma\tau_{\rm ir}/\mu} = I_0 e^{\gamma\tau_{\rm ir}/\mu}\delta(\mu+\mu_0)\delta(\phi-\phi_0). \qquad (9.55)$$

The mean intensity for an inward beam $(-1 < \mu < 0)$ from one direction μ_0 is

$$J_{\rm vis}(\tau) = \frac{1}{4\pi}\int_0^{2\pi}\int_{-1}^{0} I_0 e^{\gamma\tau_{\rm ir}/\mu}\delta(\mu+\mu_0)\delta(\phi-\phi_0)d\mu d\phi \qquad (9.56)$$

or

$$J_{\rm vis}(\tau) = \frac{1}{4\pi} I_0 e^{-\gamma\tau_{\rm ir}/\mu_0}. \qquad (9.57)$$

Integrating equation [9.53] with this $J_{\rm vis}(\tau)$, we find

$$\boxed{F_{\rm ir}(\tau) = \mu_0 I_0 e^{-\gamma\tau_{\rm ir}/\mu_0} + F_{\rm int},} \qquad (9.58)$$

where the integration constant $F_{\rm int}$ is determined by considering the upper boundary condition at $\tau_{\rm ir} = 0$, $F_{\rm ir}(0) = \mu_0 I_0 + F_{\rm int}$. Here $F_{\rm int} = \sigma_{\rm R} T_{\rm int}^4$.

With expressions for $F_{\rm ir}(\tau)$ and $J_{\rm vis}(\tau)$, we can now find an expression for $J_{\rm ir}(\tau)$ by integrating equation [9.54],

$$J_{\rm ir}(\tau) = -\frac{3}{4\pi}\left[\frac{\mu_0^2}{\gamma} I_0 e^{-\gamma\tau_{\rm ir}/\mu_0}\right] + \frac{3}{4\pi} F_{\rm int}\tau_{\rm ir} + c'. \qquad (9.59)$$

To reach a temperature profile we use $J_{\rm ir}(\tau) = \sigma_{\rm R} T^4/\pi$ and $F_{\rm int} = \sigma_{\rm R} T_{\rm int}^4$ to find

$$T^4(\tau) = \frac{3}{4} T_{\rm int}^4 \tau_{\rm ir} - \frac{3}{4}\left[\frac{1}{\sigma_{\rm R}}\frac{\mu_0^2}{\gamma} I_0 e^{-\gamma\tau_{\rm ir}/\mu_0}\right] + \frac{c'\pi}{\sigma_{\rm R}}. \qquad (9.60)$$

We may find the constant c' as before for the atmosphere heated from below, by writing down the integral for the emergent flux and solving for the constant. We also use the condition for radiative equilibrium (equation [9.48]), the equations for $J_{\rm ir}(\tau)$ and $J_{\rm vis}(\tau)$, and the upper boundary condition that the emergent flux is equal to $\mu_0 I_0 + \sigma_{\rm R} T_{\rm int}^4$. To find c' we then have to solve

$$F_{\rm ir}(0) = 2\pi \int_0^1 \int_0^\infty \left[S_{\rm ir}(\tau_{\rm ir}) e^{-\tau_{\rm ir}/\mu}\right] d\tau_{\rm ir} d\mu \equiv \sigma_{\rm R} T_{\rm int}^4 + \mu_0 I_0. \qquad (9.61)$$

Evaluating c' leads to

$$\boxed{\begin{aligned} T(\tau)^4 &= \frac{3}{4} T_{\rm int}^4 \left[\tau_{\rm ir} + 2/3\right] + \\ \mu_0 T_0^4 &\left[-\frac{3}{4}\frac{\mu_0}{\gamma} e^{-\gamma\tau_{\rm ir}/\mu_0} + \frac{3}{2}\left(\frac{2}{3} + \left(\frac{\mu_0}{\gamma}\right)^2 - \left(\frac{\mu_0}{\gamma}\right)^3 \ln\left(1 + \frac{\gamma}{\mu_0}\right)\right)\right]. \end{aligned}} \qquad (9.62)$$

Here we have associated a temperature with the incident intensity via $I_0 = \sigma_{\rm R} T_0^4$.

The temperature profile we have just derived can be used for exoplanet atmospheres. Figure 9.7 shows temperature profiles for different values of μ_0.

9.7 THE ADIABATIC TEMPERATURE PROFILE

9.7.1 Convection Basics

In a planetary atmosphere there is a net transport of energy outward. So far we have discussed energy transport by radiation (i.e., the microscopic interactions of photons with atoms and molecules). In this section we will discuss energy transport by convection.

Convection is the transport of energy by bulk motions of matter in the atmosphere in the vertical direction. During convection, heat flows from hotter to cooler regions by the macroscopic movement of matter. Convection in a planetary atmosphere is driven by the temperature gradient and is enabled because of the gravity field.

To describe an overview of convection we consider a local air parcel that is heated from below. The heated air parcel becomes less dense than its surroundings and buoyancy forces will cause the air parcel to rise. As the air parcel rises, it experiences a lower ambient pressure, and expands and releases heat. Now cooler than the surrounding atmosphere, the air parcel sinks to a level with higher pressure and higher temperature. The parcel is again heated, again rises to release heat, sinks again, and repeats this in the convective energy cycle.

Everyday examples of convection occur on many different scales and include boiling water, lava lamps, and "shimmering" air above hot pavement. Convection in the Sun's atmosphere is evident from images that show a large granulation pattern. The bright spots are centers of convective cells—the tops of columns of rising hot gas. The dark areas are the areas of cooled gas beginning to descend. The solar convective cells are large—typically 1000 km in diameter.

9.7.2 Derivation of the Adiabatic Temperature Profile

We want to derive a temperature change with altitude (a temperature profile) that occurs in a gas as it adiabatically expands or is compressed as it rises under hydrostatic equilibrium. We begin with basic thermodynamic principles; describing them in terms of temperature and pressure will lead us to a formulation of the adiabatic temperature profile.

The first law of thermodynamics is the conservation of energy applied to thermodynamic processes. Energy can be transferred from one system to another but it cannot be created or destroyed. Work done on or by a system must change the internal energy of that system. The change in internal energy dU is

$$dU = dQ - dW, \tag{9.63}$$

where dQ is the amount of energy added to the system by heating and dW is the the amount of energy lost by via work done by the system. We should consider this conservation of energy description as applied to an air parcel, and we will use this equation to derive the adiabatic temperature gradient. We will now take each term of equation [9.63] in turn and describe it in terms of T and P. From the definition of specific heat capacity and the relation between c_v and c_p we rewrite the change in internal energy dU as

$$dU = mc_v dT = m(c_p - R_s)dT, \tag{9.64}$$

where m is mass, c_v and c_p are the specific heat capacities at constant volume and constant pressure respectively, and R_s is the specific gas constant; all have units of J kg^{-1} K^{-1}. An adiabatic process has no heat exchanged with the surroundings so that $dQ = 0$. In an expanding volume work is defined as $dW = PdV$. Taking the derivative of the equation of state for an ideal gas, we obtain

$$PdV = mR_s dT - VdP = mR_s dT - \frac{m}{\rho} dP. \tag{9.65}$$

We can now rewrite the conservation of energy in equation [9.63] as

$$\frac{dT}{dP} = \frac{1}{\rho c_p}. \tag{9.66}$$

To convert this temperature-pressure relation to a temperature gradient we relate the pressure P to altitude z by using the hydrostatic equilibrium equation [9.5]

$$dP = -\rho g dz \tag{9.67}$$

to derive the adiabatic temperature gradient

$$\boxed{\Gamma \equiv -\frac{dT}{dz} = \frac{g}{c_p}.} \tag{9.68}$$

We may also write the adiabatic temperature gradient as a function of optical depth

$$\Gamma_\tau \equiv \frac{dT}{d\tau} = \frac{g}{\overline{\kappa}(\tau) c_p}, \tag{9.69}$$

where the negative sign on the right-hand side comes from the increase of the optical depth scale with decreasing altitude. The adiabatic temperature gradient is called the adiabatic lapse rate, Γ, for planets. Note the sign convention; Γ is positive for a decrease in temperature with altitude. The adiabatic lapse rate is the temperature change that occurs in the gas as it adiabatically expands or is compressed. Returning to the air parcel scenario, the adiabatic lapse rate means that an air parcel moving in a hydrostatic atmosphere has a fixed rate of change of both temperature and density with altitude. The air parcel will have its own temperature and density, as given by the adiabatic lapse rate, but will share the pressure of the surrounding atmosphere.

The above equation for the adiabatic lapse rate is for a "dry" atmosphere. In Earth's atmosphere, the release of latent heat from condensation of water vapor must be considered in the lapse rate. Although it is not strictly adiabatic, we can compute the temperature profile by considering the heat deposited in the condensing layer, $dQ = -dm_s L$, where dm_s is the change in the mass of the condensing vapor per unit mass of noncondensible gas and L is the specific latent heat for the vapor (in units of J kg^{-1}). The negative sign arises because the air parcel absorbs heat—the latent heat deposited in the layer by the condensing vapor. Keeping dQ in the first law of thermodynamics we may follow the above derivation to find the moist adiabatic lapse rate

$$\boxed{\Gamma \equiv -\frac{dT}{dz} = \frac{g}{c_p}\left[\frac{1}{1 + (L/c_p)(dm_s/dT)}\right].} \tag{9.70}$$

The difference between the dry and moist adiabatic lapse rates can be significant. For Earth's troposphere, the dry adiabatic lapse rate is about 9.8 K/km. The moist adiabatic lapse rate is about 5 K/km. Typically, an average value of 6.5 K/km can be used, because convection in Earth's trophosphere is partially moist and partially dry. Remarkably, with the dry adiabatic lapse rate Earth's atmosphere is stable against convection, and with the moist adiabatic lapse rate, convection sets in.

Many exoplanets with different atmospheric temperatures from Earth's will have condensation of gas other than water vapor. We could use the same formulation for the moist adiabatic lapse rate, substituting the relevant change in masses of saturated gas and specific latent heat values.

9.8 THE ONE-DIMENSIONAL TEMPERATURE-PRESSURE PROFILE

9.8.1 Conceptual Overview

The approach to determining the temperature profile with altitude in a planetary atmosphere in radiative-convective equilibrium is to solve a set of three equations (the radiative transfer equation, the radiative and convective equilibrium equation, and the hydrostatic equilibrium equation) for three unknowns (the temperature, pressure, and radiation field). The approach to the solution is to adopt a starting solution via a temperature pressure profile and then to iterate until a T-P profile is computed that satisfies the three equations.

For example, we could start with a T-P profile from the gray atmosphere (Section 9.6.3) using the appropriate temperature- and pressure-dependent opacities. We would then solve the radiative transfer equation, the radiative equilibrium equation, and the hydrostatic equilibrium equation by any one of the number of numerical methods available. These numerical methods typically derive temperature corrections and iterate until a T-P profile that satisfies all three equations is obtained. With the new T-P profile, we would test at each altitude whether or not the atmosphere is stable against convection. If the atmosphere were stable against convection, our solution would be completed.

If in this example the atmosphere were to be unstable, convection would be occuring and we would use the adiabatic lapse rate as the temperature profile in the convective part of the atmosphere. We would iterate again to find a T-P profile in the radiative region, and one that smoothly connects to the adiabatic temperature profile.

9.8.2 A Radiative Equilibrium Atmosphere

For an atmosphere in radiative equilibrium, we wish to solve three equations: the equation of hydrostatic equilibrium (equation [9.6]),

$$\frac{dP}{d\tau} = \frac{g\rho}{\overline{\kappa}}; \tag{9.71}$$

the equation of radiative transfer (equation [5.39]),

$$\mu \frac{dI(\tau, \mu, \nu)}{d\tau} = I(\tau, \mu, \nu) - S(\tau, \mu, \nu); \tag{9.72}$$

and the equation of radiative equilibrium (equation [9.31]).

$$\frac{dF(\tau)}{d\tau} = \int_0^\infty \kappa(\tau,\nu)\,[J(\tau,\nu) - S(\tau,\nu)]\,d\nu = 0. \tag{9.73}$$

Outputs. In these equations there are three unknowns: the radiation field $I(\tau,\mu,\nu)$, the temperature $T(\tau)$, and the pressure $P(\tau)$. The three equations can be solved simultaneously for the three unknowns; thus the atmospheric T-P profile and the radiation field emerge simultaneously.

Inputs. The additional parameters in the three vertical structure equations are the interior temperature $T_{\rm int}$ and the surface gravity g and these are classifications of the model. g is known for transiting exoplanets, and $T_{\rm int}$ comes from evolution calculations or is a free parameter of the model. The opacities are another set of inputs. The opacities depend upon the composition of the atmosphere. The elemental abundance is an input variable and chemical equilibrium is assumed or a photochemical model must be used (Chapter 4).

Subtleties. The above system of equations is highly coupled and benefits from a simultaneous solution. For example, if the radiative transfer equation were solved on its own, and the solution did not satisfy radiative equilibrium, a new temperature profile $T(\tau)$ would have to be found that did satisfy radiative equilibrium. But, for a different temperature structure, the number densities of different chemical species would change. As a result the gas pressure would change, as well as the opacities and emissivities $\kappa(\tau,\nu)$ and $\varepsilon(\tau,\mu,\nu)$, leading to a change in the radiation field $I(\tau,\mu,\nu)$ at all depths.

The radiative transfer equation itself is highly nonlinear, as the scattering term means that photons decouple the radiation field from the local temperature. For the case of exoplanets irradiated by their parent stars, the radiation field at the top of the atmosphere is coupled to the deeper atmosphere by scattering—photons can travel long distances down into the atmosphere before heating the atmosphere.

To solve the equations, iterations are usually required, beginning with an estimated temperature pressure profile, typically from a gray atmosphere solution.

9.8.3 Radiative and Convective Equilibrium

To determine whether or not convection should be occurring in a planetary atmosphere, we take a radiative equilibrium temperature profile, and check if the atmosphere is stable against convection, with the previously described criterion for convection in equation [9.23] (but replacing "atm" with "rad"),

$$\left(\frac{dT}{dz}\right)_{\rm ad} > \left(\frac{dT}{dz}\right)_{\rm rad}. \tag{9.74}$$

In planetary atmosphere models one usually assumes that if convection is occurring, convection is so efficient that it is the dominant process for energy transport. This means that the temperature gradient comes directly from the adiabatic lapse rate (equation [9.69]). In the lapse rate equation, hydrostatic equilibrium has already been included. The temperature as a function of pressure can be computed from the adiabatic lapse rate and with the ideal gas law.

The adiabatic lapse rate by itself gives the slope of the temperature gradient, but describes any of a number of "parallel" lines in temperature versus altitude (or in temperature versus pressure depth). We need to derive a temperature profile consistent with the temperature profile in the radiative equilibrium part of the planetary atmosphere. To do this we use as a boundary condition the radiative equilibrium temperature at the upper end of the convection zone. Here we are assuming that convective equilibrium holds, that is,

$$F_{\text{conv}}(\tau) = \sigma_{\text{R}} T_{\text{int}}^4 + \mu_0 I_0. \tag{9.75}$$

There may be a region in a real atmosphere where some of the energy is transported by radiation and the rest by convection. It is complicated to solve explicitly for the fraction of the radiation transported by convection. To proceed we would substitute radiative and convective equilibrium

$$F_{\text{rad}}(\tau) + F_{\text{conv}}(\tau) = \sigma_{\text{R}} T_{\text{eff}}^4 + \mu_0 I_0 \tag{9.76}$$

for the radiative equilibrium in equations [9.71]–[9.73] above. The major complication arises in developing an expression for the convective flux. Indeed, only an approximate expression is derivable, and here we will provide only an overview.

In order to determine how much energy is transported by convection compared to radiation, we will need to know the convective flux. Recall from Section 9.7.1 that during convection, heat flows from hotter to cooler regions by the macroscopic movement of matter. The excess energy deposited per unit volume when a mass element merges with the surrounding atmosphere is $\rho c_p \Delta T$. The heat flux is then

$$F_{\text{conv}} = \bar{v} \rho c_p \Delta T, \tag{9.77}$$

where $\bar{v}$ is the mean velocity of the mass element. Here ΔT arises from the difference between the temperature gradient of the rising material and the temperature gradient of the surrounding atmosphere. Following [8], ΔT can be expressed in terms of the temperature gradients for mass elements moving over the distance Δz,

$$\Delta T = \left[\left(-\frac{dT}{dz} \right) - \left(-\frac{dT}{dz} \right)_{\text{c}} \right] \Delta z, \tag{9.78}$$

where the first temperature gradient on the right hand side is that of the surrounding atmosphere in the sought after state of radiative and convective equilibrium. The second temperature gradient is that of the convective elements.

Let us take a look at the above two equations to see what we would need to derive the convective flux. The heat capacity should be known, the temperature is what we are trying to solve for, and the density comes from the ideal gas via the hydrostatic equilibrium equation. We are left with the convective velocity and the distance convective elements travel before releasing heat to the surroundings.

The derivation of an expression for the convective velocity, the temperature gradient, and the distance a convective element travels is complex with many caveats and takes several pages. We leave the details to the excellent references [6,7,8]. These references are to stellar atmospheres, where the theory of convection is a 1D approximation called the mixing length theory. This theory is a local theory because no definitive 3D convective theory is available. The main assumption in the mixing length theory is the scale length l, the distance over which a convective bubble rises and releases its heat, before sinking again. The choice of l is somewhat arbitrary, but it is approximately comparable to the pressure scale height.

9.9 TEMPERATURE RETRIEVAL

Up until now we have described the so-called forward problem, starting with basic physics to derive a planet's vertical temperature structure. The forward problem involves solution of a set of coupled equations, beginning with a starting solution and iterating to convergence. A direct solution for the temperature, or the "inverse problem," is possible if a highly detailed spectrum is available. Studies of solar system planetary atmospheres use temperature retrieval methods by first inferring a basic T-P profile and then perturbing that fiducial temperature profile until data is fit well. The fiducial temperature profile is perturbed over both temperature and molecular abundances (see, e.g., [1]).

For exoplanets—in contrast to solar system planets—the spectra are not of sufficient quality to infer a unique fiducial temperature-pressure profile (see Figures 6.6, 6.7, and 6.8). In other words, there is no starting point from the data to derive a fiducial model. A suitable approach is to computationally derive the *range* of temperature-pressure profiles and molecular and atomic abundances allowed by a given spectrum. The forward temperature-pressure structure determination described in this chapter, however, is too computationally intensive to use to run millions of models to find a good fit.

A new method for exoplanet temperature and abundance retrieval has therefore been proposed [9]. This new method uses a parametric T-P profile (Figure 9.8), running tens of thousands of them that fit the data, and requiring T-P profiles to satisfy hydrostatic equilibrium and global energy balance. The quantitatively allowed ranges of T-P profiles and molecular abundances can be described. In addition, constraints on the albedo and day-night energy redistribution and on the effective temperature can be determined.

The parametric T-P profile in this example is motivated by physical principles, solar system planet T-P profiles, and 1D exoplanet T-P profiles generated from model atmosphere calculations reported in the literature. The P-T parametric profile is a generalized exponential profile of the form

$$P = P_0 e^{\alpha(T-T_0)^{\beta}}, \tag{9.79}$$

where P is the pressure in bars, T is the temperature in K, and P_0, T_0, α, and β are free parameters. For layer 3, the model profile is given by $T = T_3$, where T_3 is a free parameter. Furthermore,

$$\begin{aligned} &P_0 < P < P_1: && P = P_0 e^{\alpha_1(T-T_0)^{\beta_1}} && \text{layer1} \\ &P_1 < P < P_3: && P = P_2 e^{\alpha_2(T-T_2)^{\beta_2}} && \text{layer2} \\ &P > P_3: && T = T_3 && \text{layer3} \end{aligned} \tag{9.80}$$

and P is typically used as the independent coordinate. The above parametric profile is in fact overspecified. The β parameter turns out to be a redundant parameter and can be set $\beta_1 = \beta_2 = 0.5$. Then the model profile in (9.80) has nine parameters, namely, P_0, T_0, α_1, P_1, P_2, T_2, α_2, P_3, and T_3. Furthermore, two of the parameters can be eliminated based on the two constraints of continuity at the two layer boundaries, that is, layers 1–2 and layers 2–3. P_0 is generally set to $P_0 = 10^{-5}$ bar,

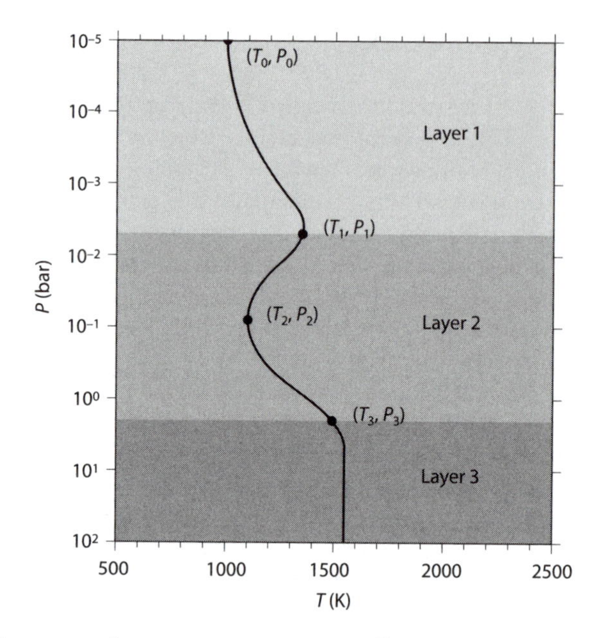

Figure 9.8 The parametric pressure-temperature profile. In the general form, the profile includes a thermal inversion layer (layer 2) and has six free parameters. An isothermal profile is assumed below the pressure P_3 (layer 3). Alternatively, for cooler atmospheres with no isothermal layer, layer 2 could extend to deeper layers and layer 3 could be absent. Adapted from [9].

that is, at the top of the model atmosphere. The parametric profile in its complete generality therefore has six free parameters.

The parametric T-P profile shown in Figure 9.8 is motivated by the physics that sets a planet atmosphere's vertical structure. In general the temperature structure at a given altitude depends on the opacity at that altitude, along with density and gravity.

The T-P profile shown in Figure 9.8 is divided into three representative layers. Above layer 1, at pressures below $P \sim 10^{-5}$ bar, the optical depth at all wavelengths becomes low enough so that the layers of the atmosphere are transparent to the incoming and outgoing radiation and not relevant for spectral features. The uppermost layer, layer 1, has no thermal inversions. Here the atmosphere is being heated by lower layers and cools with increasing altitude. The middle layer, layer 2, is where most spectral features are formed. In layer 2, the temperature structure is governed by radiative process and possibly by atmospheric dynamics. These optically thin layers are at altitudes where thermal inversions may be formed, depending on the level of irradiation from the parent star and the presence of strong absorbing gases or solid particles (see Figure 9.1). The bottom-most layer, layer 3, is the regime where a high optical depth leads to radiative diffusion and the related isothermal temperature structure. Essentially the strong irradiation heating from above does not reach the deep atmosphere layer, which is heated from the interior energy outflow. Below layer 3, in the deepest layers of the planet atmosphere, convection is the dominant energy transport mechanism. The high pressure (equivalently, high density) implies a high opacity, making energy transport by convection a more efficient energy transport mechanism than radiation.

The description of the T-P profile in Figure 9.8 is focused on hot Jupiters. For cooler planets layer 3 could be absent, with layer 2 extending to deeper layers. In addition, the radiative-convective boundary occurs at a higher altitude in the planet atmosphere than for hot Jupiters, meaning convection may play a role in layers 2 and 3, making the temperature profile an adiabat.

9.10 SUMMARY

The vertical temperature structure of a planetary atmosphere is intimately connected to energy conservation: energy is neither created nor destroyed in an atmospheric layer. We began with a description of Earth's vertical atmospheric structure. We continued with a derivation of the hydrostatic equilibrium equation, which describes how atmospheric pressure supports the atmosphere against gravity. The vertical thermal structure is also connected to energy transport. We described the two major mechanisms of energy transport, radiation and convection, and the criterion that determines which energy transport mechanism dominates. If energy transport by radiation dominates, then the vertical thermal structure comes from radiative equilibrium. If convection dominates, the vertical thermal structure is defined by the adiabatic lapse rate.

The planetary spectrum can be connected to the vertical thermal structure—that is, the temperature gradient—by energy conservation, hydrostatic equilibrium, radiative transfer, and opacities. The 1D temperature structure (by which we really mean the temperature-pressure structure) and the emergent planetary spectrum can be solved with three equations (radiative transfer, radiative and convective equilibrium, and hydrostatic equilibrium) for three unknowns as a function of altitude (the temperature, pressure, and frequency-dependent radiation field).

Ultimately the vertical temperature-pressure structure is tied to the opacities and radiative transfer: if energy transport is by radiation, the opacity governs the temperature structure. If energy transport by radiation is inhibited, then convection takes over.

REFERENCES

For further reading

For a description of convection, including mixing length theory:

- Bohm-Vitense, E. 1989. *Stellar Atmospheres, vol. 2 and 3* (Cambridge: Cambridge University Press).

- Hansen, J., Kawaler, S. D., and Trimble, V. 2004. *Stellar Interiors: Physical Principles, Structure, and Evolution* (2nd ed; New York: Springer).

For a basic description of convection as applied to Earth's atmosphere:

- Marshall, J., and Plumb, A. 2008. *Atmosphere, Ocean, and Climate Dynamics: An Introductory Text* (London: Elsevier Academic Press).

References for this chapter

1. Goody, R. M., and Yung, Y. L. 1989. *Atmospheric Radiation: Theoretical Basis* (2nd ed; Oxford: Oxford University Press).

2. U.S. Standard Atmosphere 1976, U.S. Government Printing Office, Washington, D.C.

3. Marshall, J., and Plumb, A. 2008. *Atmosphere, Ocean, and Climate Dynamics: An Introductory Text* (London: Elsevier Academic Press).

4. Guillot, T., Gautier, D., Chabrier, G., and Mosser, B. 1994. "Are the Giant Planets Fully Convective?" Icarus 112, 337–353.

5. Hansen, B. M. S. 2008. "On the Absorption and Redistribution of Energy in Irradiated Planets." Astrophys. J. 179, 484–508.

6. Bohm-Vitense, E. 1989. *Stellar Atmospheres, vol. 2 and 3* (Cambridge: Cambridge University Press).

7. Hansen, J., Kawaler, S. D., and Trimble, V. 2004. *Stellar Interiors: Physical Principles, Structure, and Evolution* (2nd ed; New York: Springer).

8. Mihalas D. 1978. *Stellar Atmospheres* (2nd ed.; San Francisco: W. H. Freeman).

9. Madhusudhan, N., and Seager, S. 2009. "A Temperature and Abundance Retrieval Method for Exoplanet Atmospheres." Astrophys. J. 707, 24–39.

EXERCISES

1. Rewrite the ideal gas law in a form that uses the universal gas constant R.

2. What is the value of the scale height H for exoplanets? Consider a hot Jupiter, a hot super Earth with an H_2-dominated atmosphere, a hot super Earth dominated by a CO_2 atmosphere, and Earth itself. Use equation [9.11].

3. Show that for an n-layer leaky greenhouse model, $T_s = (n+1)^{1/4} T_e$, where T_e is the emission temperature in the highest atmosphere layer.

4. Complete the derivation of the gray, radiative equilibrium, LTE, no-scattering atmosphere.

   a. Derive $3c = \frac{\sigma_{\rm R} T_{\rm int}^4}{2\pi}$ by solution of the equation [9.44].

   b. Do the same for equation [9.62].

5. Assumptions in the gray radiative equilibrum temperature profile.

   a. In the derivation of the gray radiative equilibrum temperature profile, we have used the Eddington approximation $J = 3K$. What limitations does this assumption put on the temperature profile?

   b. We derived the same temperature gradient from the gray radiative equilibrum temperature profile (equation [9.45]) as was derived in the diffusion approximation (Section 6.4.4, equation [6.53]). What assumptions that went into the gray radiative equilibrium temperature profile cause this to be the same?

6. Estimate the surface pressure of an exoplanet, given an atmospheric mass and composition.

7. Derive the equation for the moist adiabatic lapse rate, equation [9.70].

8. The solar system planet atmosphere vertical temperature-pressure profiles shown in Figure 9.1 have remarkable similarity because temperature profiles of planetary atmospheres are governed by basic physics. Qualitatively describe the physics that causes the characteristic shape of the temperature-pressure profiles.

Chapter Ten

Atmospheric Circulation

10.1 INTRODUCTION

Atmospheric circulation is the large-scale movement of gas in a planetary atmosphere that is responsible for distributing energy absorbed from the star throughout the planetary atmosphere. Curiously, many textbooks on atmospheric radiation omit discussion of atmospheric circulation. Conversely, textbooks on atmospheric circulation often relegate a description of radiation in the atmosphere to one chapter or less. This segregation happens for two reasons. First, for solar system planets, interpretation of spectra in terms of molecular abundances and vertical temperature structure often does not require atmospheric circulation. Second, calculation of atmospheric circulation and dynamics is computationally time consuming, and the timescales of atmospheric dynamics are very different from those of radiation. Atmospheric circulation codes can usually only afford to have a relatively crude radiation scheme.

Atmospheric dynamics focuses on observational and theoretical analysis of motion in the atmosphere on a diverse scale. On Earth, atmospheric circulation models are used both to predict weather and to assess climate change. Atmospheric circulation is also responsible for many large- and small-scale phenomena such as the trade winds, the jet stream, the El Niño and La Niña weather oscillations, hurricanes and tornadoes, and the jet stream. On Jupiter, atmospheric circulation causes high wind speeds in jets and the intricate weather patterns on Jupiter, including the great red spot.

A very fundamental issue is that on all of the solar system planets atmospheric circulation acts to minimize temperature gradients, in part by transporting heat poleward. This means that, despite all of the atmospheric phenomena on all planetary atmospheres created by atmospheric circulation, the longitudinal and latitudinal temperature variation is relatively small, as are the resultant emergent spectra. (One notable exception is a hemispherically integrated spectrum of Earth centered on Earth's cold poles.) Figure 10.1 shows that the latitudinal temperature varies for less than a few degrees for the solar system giant planets. For these planets the 1D temperature profile approach described in (Chapter 9) and earlier chapters is adequate to infer the vertical temperature gradient, surface temperature, and atmospheric composition.

On some planets beyond our solar system, such as hot tidally locked exoplanets, atmospheric dynamics does affect the emergent spectra. We then have a motivation to study atmospheric circulation beyond basic planetary knowledge to use atmospheric dynamics for interpreting spectra. The hot Jupiters are several times closer

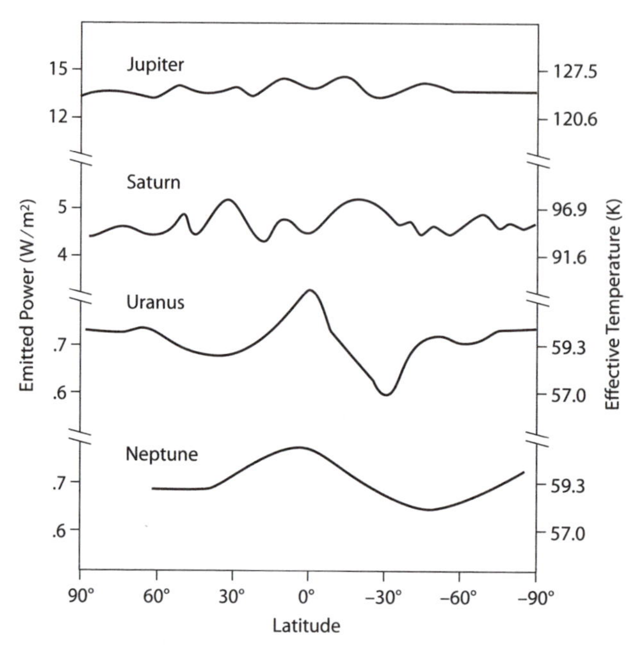

Figure 10.1 Temperature versus latitude on solar system giant planets. The solar system giant planets show almost no temperature change with latitude. Adapted from [1].

to their star than Mercury is to our Sun, receiving 400 times more radiation than Earth and 10,000 times more than Jupiter. The energy from stellar heating dominates any internal flux for hot Jupiters. Together with the permanent day and night sides, this case of stellar forcing has no solar system analog. The fate of the absorbed stellar energy and the global temperature structure of the planet can only be understood via atmospheric circulation. The question of energy redistribution is especially significant for habitable-zone planets orbiting low-mass M stars. Like their hot-Jupiter counterparts, such planets are so close to the star that they are tidally locked, presenting the same face to the star at all times. Whether or not the planet is habitable depends on the role played by atmospheric circulation. By "tidal locking" we are referring to the planet core; an additional complication is to what extent the atmosphere departs from tidal locking. A reasonable picture for planets with thick atmospheres is a tidally locked core with a thick mobile atmosphere.

To complete the picture of the thermal structure and emergent flux characteristics of the tidally locked exoplanet atmospheres, the dynamical response of the atmosphere to heat sources and sinks is the final major process we must consider. These considerations are most significant for tidally locked exoplanets. Due to the complexity and nonlinearity of the atmospheric circulation equations, the full derivation and application of the equations are beyond the scope of this book. Similarly, because of the complexity of the equations, an intuitive, conceptual understanding of atmospheric dynamics is often elusive. In this chapter we will focus mostly on

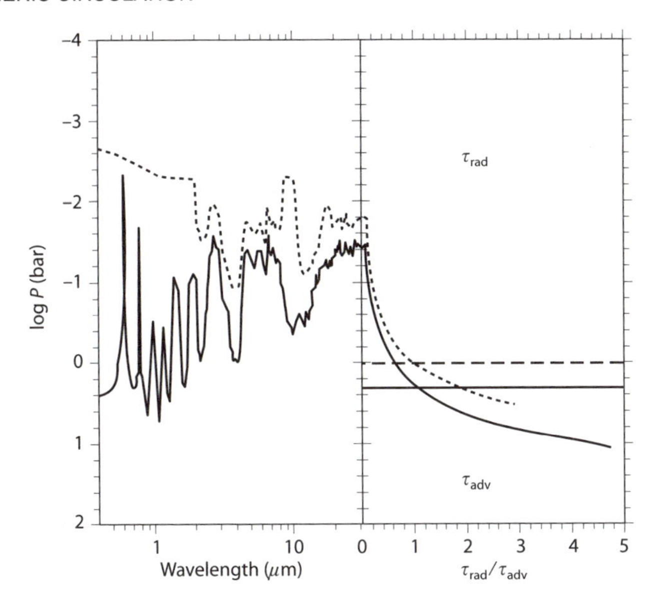

Figure 10.2 Illustration of $\tau_{\rm rad}$ vs. $\tau_{\rm adv}$. Left: pressure (as a proxy for altitude) at optical depth of 2/3 as a function of wavelength. Two different models are shown (cloudy by the dashed curve and clear by the solid curve). Right: altitude dependence of the ratio of the radiative to advective timescales ($\tau_{\rm rad}/\tau_{\rm adv}$). A wind speed U of 1000 m s^{-1} was adopted for illustration; the ratio scales linearly with U so that other values can be considered. Adapted from [3].

timescales and parameters that describe large-scale flow. We will also present a schematic outline of the equations for atmospheric circulation.

10.2 RADIATIVE AND ADVECTIVE TIMESCALES

We begin by estimating the typical timescales that govern whether or not atmospheric circulation is important in causing a longitudinal or latitudinal temperature gradient. This discussion is necessarily oversimplified but helps to illustrate some basic points. Essentially, there is a competition between the radiative and the advective timescales. The radiative timescale ($\tau_{\rm rad}$) is the time for absorbed stellar energy to be reemitted as radiation. The advective timescale ($\tau_{\rm adv}$) is the time for the absorbed stellar energy to be circulated around the planet.

If $\tau_{\rm rad} \ll \tau_{\rm adv}$, the bulk of the absorbed stellar energy will be reemitted before being advected around the planet. In this case, a strong latitudinal and longitudinal temperature gradient are expected to arise. If, in contrast, $\tau_{\rm adv} \ll \tau_{\rm rad}$, the temperature should be much more uniform, because heat would be transported and redistributed efficiently over the entire planet.

The radiative timescale can be estimated by considering an atmosphere layer of thickness Δz that slightly perturbed from radiative equilibrium. Consider an

Table 10.1 Comparison of radiative and advective timescales of some solar system planets.

Planet	$\tau_{\rm rad}$	$\tau_{\rm adv,lon}$	$\Delta T_{\rm lon}$	$\tau_{\rm adv,lat}$	$\Delta T_{\rm lat}$
Venus	years	days	$<$ few K	weeks	few K
Earth	weeks–months	1 day	$\sim$10 K	weeks	20–30 K
Mars	days	1 day	$\sim$50 K	weeks	$\sim$100 K
Jupiter	years–decades	decades	$<$ few K	decades	$<$ few K

Estimated temperature differences in longitude and latitude are also presented. Table adapted from [2] and references therein.

atmospheric layer of pressure thickness ΔP that is out of radiative equilibrium by an amount ΔT. This layer has excess energy per area $\rho c_p \Delta T \Delta z$. Approximating the radiation as black body, the layer will radiate a net excess of $4\sigma_{\rm R} T^3 dT$. Taking the above two statements and using the hydrostatic equilibrium equation gives

$$\boxed{\tau_{\rm rad} \sim \frac{\Delta P}{g} \frac{c_p}{4\sigma_{\rm R} T^3}.} \tag{10.1}$$

Here T is temperature, g is surface gravity, c_p is the specific heat capacity, and $\sigma_{\rm R}$ is the radiation constant. P/g is equivalent to mass/area. This estimate applies only to regions of low optical depth. The advective timescale can be estimated by the planet radius and the characteristic windspeed U,

$$\boxed{\tau_{\rm adv} \sim \frac{R_{\rm p}}{U}.} \tag{10.2}$$

Here U is unknown for exoplanets—indeed the windspeed is a key term one wants to derive from atmospheric circulation models. We now use these timescale estimates to investigate temperature contrasts on planets, at altitudes where the bulk of solar energy is absorbed. We can see from Table 10.1 that, indeed, a longer $\tau_{\rm rad}$ leads to a smaller latitudinal and longitudinal temperature gradient. Although the advective timescales on solar system giant planets such as Jupiter are long, the radiative times are even longer. Furthermore, the fast rotation rate ($\sim$12 hours; see Table A.1) in part means that Jupiter is heated relatively uniformly.

We now turn to an important subtlety in our radiative versus advective timescale analysis. The ratio of the radiative-to-advective timescales depends on the vertical altitude where the stellar radiation is absorbed. This is primarily because $\tau_{\rm rad}$ decreases rapidly deeper into the atmosphere; it depends linearly on pressure, which increases exponentially deeper into the atmosphere (section 9.3.1). In general, the above description of competing timescales is valid for the altitude where the bulk of the stellar energy is absorbed.

Different layers of the atmosphere may have different circulation regimes. Earth, for example, has a relatively small latitudinal and longitudinal temperature difference in the troposphere where we live. High in the thermosphere, however, the day-night temperature difference can be as high as 1000 K. How can we observe different atmospheric layers on exoplanets? Recall that the optical depth is frequency dependent; molecules absorb more strongly at specific frequencies. We explained in previous chapters that this means we can see different layers of the atmosphere depending on the frequency of observation. Figure 10.2 shows the optical depth unity for a hot Jupiter atmosphere to illustrate that optical depth—and hence radiative versus advective regimes—are at different altitudes for different wavelengths.

Can we repeat the radiative versus advective timescale analysis for exoplanets? The main issue in understanding which regime the atmosphere is in is the unknown windspeed U. This value is not known on exoplanets but can either be estimated (in a circular fashion by using $\tau_{\rm rad}$ vs. $\tau_{\rm adv}$ and an estimated temperature) or obtained from computer simulations.

10.3 LARGE-SCALE FLOW AND PATTERNS

Characteristic length scales can tell us something about the big picture of heat transport and weather patterns on a given exoplanet. Before describing two major characteristic scales, we discuss the transformation from an inertial to a rotating reference frame.

10.3.1 The Rotating Reference Frame

In planetary atmospheric circulation it is natural to use the rotating frame of reference, where the so-called inertial or fictitious forces appear. These are the centrifugal and Coriolis forces. Here we will present an outline of the derivation of a transformation from the inertial to the rotating reference frame, that essentially results in the addition of the centrifugal and Coriolos "forces." We will follow [4, 5] in our derivation. Let us take $\mathbf{A}$ as an arbitrary vector in a Cartesian inertial reference frame. The vector is described by

$$\mathbf{A} = A_x\hat{\mathbf{i}} + A_j\hat{\mathbf{j}} + A_k\hat{\mathbf{k}}, \tag{10.3}$$

where $\hat{\mathbf{i}}$, $\hat{\mathbf{j}}$, and $\hat{\mathbf{k}}$ are unit vectors. We will describe the same vector in a reference frame rotating with angular velocity $\mathbf{\Omega}$.

$$\mathbf{A} = A'_x\hat{\mathbf{i}}' + A'_j\hat{\mathbf{j}}' + A'_k\hat{\mathbf{k}}'. \tag{10.4}$$

The total Lagrangian derivative of $\mathbf{A}$ in the inertial frame (subscript I) is

$$\begin{aligned}\frac{D_I\mathbf{A}}{Dt} &= \frac{DA_x}{Dt}\hat{\mathbf{i}} + \frac{DA_y}{Dt}\hat{\mathbf{j}} + \frac{DA_k}{Dt}\hat{\mathbf{k}} \\ &= \frac{DA'_x}{Dt}\hat{\mathbf{i}}' + \frac{DA'_y}{Dt}\hat{\mathbf{j}}' + \frac{DA'_k}{Dt}\hat{\mathbf{k}}' + \frac{D_I\hat{\mathbf{i}}'}{Dt}A'_x + \frac{D_I\hat{\mathbf{j}}'}{Dt}A'_y + \frac{D_I\hat{\mathbf{k}}'}{Dt}A'_z.\end{aligned} \tag{10.5}$$

We can use the definition of the derivative in the rotating frame

$$\frac{D\mathbf{A}}{Dt} = \frac{DA'_x}{Dt}\hat{\mathbf{i}}' + \frac{DA'_y}{Dt}\hat{\mathbf{j}}' + \frac{DA'_k}{Dt}\hat{\mathbf{k}}' \tag{10.6}$$

and the cross product terms for the velocity of the unit vectors caused by rotation, e.g.,

$$\frac{D_I\hat{\mathbf{i}}'}{Dt} = \mathbf{\Omega} \times \hat{\mathbf{i}}', \tag{10.7}$$

to write equation [10.5] as

$$\boxed{\frac{D_I\mathbf{A}}{Dt} = \frac{D\mathbf{A}}{Dt} + \mathbf{\Omega} \times \mathbf{A}.} \tag{10.8}$$

We can similarly derive a transformation for the rate of change of velocity $\mathbf{U}$ in the rotating frame of reference (again following [4]). We want to find an expression for $\frac{D_I\mathbf{U}_I}{Dt}$. We may use a position vector $\mathbf{r}$ to write

$$\frac{D_I\mathbf{r}}{Dt} = \mathbf{U}_I. \tag{10.9}$$

Applying our transformation equation [10.8] we have

$$\frac{D_I\mathbf{r}}{Dt} = \frac{D\mathbf{r}}{Dt} + \mathbf{\Omega} \times \mathbf{r} \tag{10.10}$$

or

$$\mathbf{U}_I = \mathbf{U} + \mathbf{\Omega} \times \mathbf{r}. \tag{10.11}$$

We may now again apply our transformation equation [10.8] to $\mathbf{U}_I$ to find

$$\frac{D_I\mathbf{U}_I}{Dt} = \frac{D\mathbf{U}_I}{Dt} + \mathbf{\Omega} \times \mathbf{U}_I. \tag{10.12}$$

This equation can be worked out to

$$\boxed{\frac{D_I\mathbf{U}_I}{Dt} = \frac{D\mathbf{U}}{Dt} + 2\mathbf{\Omega} \times \mathbf{U} - \Omega^2\mathbf{R},} \tag{10.13}$$

where the details are left as an exercise. Here $\mathbf{\Omega}$ is taken to be a constant and $\mathbf{R}$ is a vector perpendicular to the axis of rotation. The magnitude of $\mathbf{R}$ is equal to the distance to the axis of rotation.

The second term on the right-hand side of equation [10.13] is known as the Coriolis force or the Coriolis acceleration and the second term is the centrifugal force. Overall, equation [10.13] tells us that the acceleration following the motion in an inertial frame is also described in the rotating frame as the rate of change of relative velocity; and the Coriolis acceleration due to the relative motion; and the centripetal acceleration caused by the rotation of the coordinates [4].

10.3.2 The Rossby Number: Rotation

The Rossby number can tell us whether or not planetary rotation is important for a given phenomenon. Let us take a motion with a length scale L and consider the speed U. We would say the rotation of the planet is important if

$$\frac{L}{U} \geq \frac{1}{\Omega} \tag{10.14}$$

where Ω is the angular velocity of planetary rotation. Rotation is more important at high latitudes than at latitudes near the equator. We therefore replace Ω with the term

$$f = 2\Omega \sin\theta, \tag{10.15}$$

where θ is latitude and f is the Coriolis frequency. The Rossby number is

$$\boxed{R_0 = \frac{U}{Lf}.} \tag{10.16}$$

The Rossby number indicates the importance of rotation on the flow. A small Rossby number (taken to be < 1) means that the effects of planetary rotation are large compared to the fluid motions, and a large Rossby number means the opposite. Let us take the length scale of a planet as $R_{\rm p}$ and consider a Jupiter-size planet with $R_{\rm p} = 7.14 \times 10^7$ m. Jupiter's rotation rate gives a small Rossby number. A hot Jupiter synchronously rotating gives a large number.

The Rossby number depends on L and U. What appears to be small scale from an exoplanet view, such as a 100 m region in an ocean, may still have a small Rossby number. For exoplanets, we are interested in much larger scales only.

10.3.3 The Burger Number: Vertical Stratification

The Burger number is a dimensionless number that indicates the vertical stratification of a fluid, in this case the atmosphere. The Burger number is defined as [e.g., 5]

$$\boxed{\mathrm{Bu} \equiv \left(\frac{L_D}{L}\right)^2,} \tag{10.17}$$

where

$$L_D \equiv \frac{\sqrt{gH}}{|f|}, \tag{10.18}$$

and U, L, and H are characteristic velocity, length, and layer thickness scales, respectively. L_D is the Rossby deformation radius. The Rossby deformation radius is the horizontal length scale at which pressure perturbations are resisted by the Coriolis force.

A Burger number of zero indicates a flow dominated by rotation, whereas a Burger number near 2 indicates a flow dominated by stratification.

10.3.4 The Rhines Scale: Number of Bands

The Rhines scale [5] is the scale at which planetary rotation causes east-west elongation (jets),

$$L_\beta = \pi\sqrt{\left(\frac{2U}{\beta}\right)} \tag{10.19}$$

where U is the eddy windspeed and β is the latitudinal gradient of f,

$$\beta = 2\Omega\frac{\cos\phi}{R_{\rm p}}. \tag{10.20}$$

The number of bands, N, on a planet might be approximately described by [6]

$$N \sim \frac{\pi R_{\rm p}}{L_\beta}. \tag{10.21}$$

Fluid motions are free to grow in the longitudinal direction, but are confined in the latitudinal direction by the characteristic Rhines scale—in other words, by the gradient of the Coriolis force. The bands confine clouds, and may affect the cloud structures and patterns we can see in a hemispherically integrated exoplanet spectrum.

10.4 ATMOSPHERIC DYNAMICS EQUATIONS

The fundamental equations that govern atmospheric motion come from the conservation laws for momentum, mass, and energy—laws of basic physics. Typically, the assumption of hydrostatic equilibrium (Section 9.3.1) is used to remove the vertical dimension of the momentum conservation equation. An equation of state, taken as the ideal gas law, is also needed. Many textbooks are devoted to the derivation and application of the atmospheric fluid dynamic equations. We aim here to present an outline of derivation of the equations used for exoplanet atmospheric circulation and some of the approximations that lead to models commonly used in exoplanet studies. For our summary outline, we closely follow the introduction to the meterological equations in [4, 7].

We consider an infinitesimal control volume fixed in an Eulerian reference frame in the atmosphere. This is a parallelepiped with a fixed position relative to the moving atmosphere. In this infinitesimal control volume, we will account for conservation of mass, momentum, and energy.

The conservation laws that lead to the atmospheric equations involve the rates of change (i.e., derivatives) of the momenum, density, and energy. Because the atmosphere is moving, the conservation laws must deal with a moving fluid. We are using a fixed volume, and therefore our first step is to relate the local derivative at a fixed point to the "total" derivative to the total rate of change of a variable of interest. We leave it as an exercise to show that

$$\frac{D\mathbf{A}}{Dt} = \frac{\partial \mathbf{A}}{\partial t} + \mathbf{U} \cdot \nabla \mathbf{A}, \tag{10.22}$$

where $\mathbf{A}$ is a vector function of interest ($\mathbf{A} = A_x\hat{\mathbf{i}} + A_y\hat{\mathbf{j}} + A_z\hat{\mathbf{k}}$, where $\hat{\mathbf{i}}, \hat{\mathbf{j}}$, and $\hat{\mathbf{k}}$ are unit vectors). This equation tells us that the total rate of change of $\mathbf{A}$ as the particles move through some velocity field $\mathbf{U}$ is the sum of the local rate of change and the rate of change of $\mathbf{A}$ following the motion. The velocity field $\mathbf{U}$ is

$$\mathbf{U} = u\hat{\mathbf{i}} + v\hat{\mathbf{j}} + w\hat{\mathbf{k}}, \tag{10.23}$$

where u, v, and w are conventional notation in atmospheric fluid dynamics, and the unit vectors $\hat{\mathbf{i}}, \hat{\mathbf{j}}$, and $\hat{\mathbf{k}}$ are conventionally taken to be directed eastward, northward, and upward, respectively.

Remember that we are considering an infinetesimal control volume fixed relative to the moving atmosphere. The derivative expression of interest, therefore (from equation [10.22]) is

$$\boxed{\frac{\partial \mathbf{A}}{\partial t} = \frac{D\mathbf{A}}{Dt} - \mathbf{U} \cdot \nabla \mathbf{A}.} \tag{10.24}$$

10.4.1 Conservation of Momentum

The conservation of momentum begins with the familiar Newton's First Law, in an interial reference frame,

$$\mathbf{F} \equiv \mathbf{f}/m = a = \frac{D_I \mathbf{U}_I}{Dt}, \tag{10.25}$$

where f is the net force, m is the mass, $\mathbf{F}$ is the net force per unit mass (not to be confused with the radiative or convective flux described in previous chapters), $\mathbf{U}$ is the velocity field, and t is time as before. First, recall the expression for $D_I \mathbf{U}_I / Dt$ given in equation [10.13].

Second, we will describe the forces $\mathbf{F}$ on the atmosphere. We have previously discussed the forces in a 1D atmosphere, namely, the forces in the vertical direction as a balance of the pressure gradient and gravity (Section 9.3.1). In 3D the pressure gradient force is a force acting in three dimensions,

$$\nabla P = \frac{\partial p}{\partial x}\hat{\mathbf{i}} + \frac{\partial p}{\partial y}\hat{\mathbf{j}} + \frac{\partial p}{\partial z}\hat{\mathbf{k}}. \tag{10.26}$$

As an acceleration, the pressure force is written $\nabla P/\rho$. In atmospheric fluid dynamics, the gravitational force is written as an effective gravity term that includes the centripetal acceleration term

$$\mathbf{g}_{\text{eff}} = -g\hat{\mathbf{k}} - \Omega^2 \mathbf{R}. \tag{10.27}$$

The remaining forces in the atmosphere are lumped together as frictional forces per unit mass, which we will denote as $\mathbf{F}_{\mathrm{f}}$.

For the conservation of momentum, we have the equation

$$\boxed{\frac{D\mathbf{U}}{Dt} = 2\boldsymbol{\Omega} \times \mathbf{U} - \frac{1}{\rho}\nabla P + \mathbf{g} + \mathbf{F}_{\mathrm{f}}.} \tag{10.28}$$

This is the three-dimensional equation for the conservation of momentum.

We are not completely finished working with the conservation of momentum equations. For further analysis, we further expand the equations into their scalar form using a spherical coordinate system. The complication is in relating the Cartesian coordinate system (x, y, z) in our infinitesimal control volume to the spherical coordinate system (λ, ϕ, z) fixed to the Earth's reference frame, where ϕ is the latitude, λ is the longitude, and z is the vertical distance above the surface of Earth. The complication lies in that the (x, y, z) coordinate system is a function of location and therefore the unit vectors $\hat{\mathbf{i}}, \hat{\mathbf{j}}, \hat{\mathbf{k}}$ are also changing with location on Earth's surface. To continue we will adopt relationships between the coordinate systems, and between the unit vectors and latitude, longitude, and vertical direction above the surface, and take a instead of r in a thin shell approximation,

$$\begin{aligned} u &\equiv a\cos\phi \frac{D\lambda}{Dt}, \\ v &\equiv a\frac{D\phi}{Dt}, \\ w &\equiv \frac{Dz}{Dt}, \end{aligned} \tag{10.29}$$

and

$$\frac{D\mathbf{U}}{Dt} = \left(\frac{Du}{Dt} - \frac{uv\tan\phi}{a} + \frac{uw}{a}\right)\hat{\mathbf{i}} + \left(\frac{Dv}{Dt} + \frac{u^2\tan\phi}{a} + \frac{vw}{a}\right)\hat{\mathbf{j}} + \left(\frac{Dw}{Dt} - \frac{u^2+v^2}{a}\right)\hat{\mathbf{k}}. \tag{10.30}$$

The interested reader can derive these by consulting the Figures in [4], and other standard atmospheric dynamics textbooks.

We may next expand the Coriolis force by using the definition of the vector cross product and considering $\mathbf{\Omega}$ in terms of the unit vectors: $\mathbf{\Omega} = 0\hat{\mathbf{i}} + 2\Omega\cos\phi\hat{\mathbf{j}} + 2\Omega\sin\phi\hat{\mathbf{k}}$; in other words, the angular velocity has no component parallel to the unit vector $\hat{\mathbf{i}}$:

$$-2\mathbf{\Omega}\times\mathbf{U} = (2\Omega w\cos\phi - 2\Omega v\sin\phi)\,\hat{\mathbf{i}} - 2\Omega u\sin\phi\hat{\mathbf{j}} + 2\Omega u\cos\phi\hat{\mathbf{k}}. \tag{10.31}$$

We finally write an expression for $\mathbf{F}_{\mathrm{f}}$ as

$$\mathbf{F}_{\mathrm{f}} = F_{\mathrm{f}x}\hat{\mathbf{i}} + F_{\mathrm{f}y}\hat{\mathbf{j}} + F_{\mathrm{f}z}\hat{\mathbf{k}}. \tag{10.32}$$

We now take the above three equations, as well as our vector notation expressions for ∇P (equation [10.26]) and $\mathbf{g}_{\mathrm{eff}}$ (equation [10.27]). We substitute these equations into the conservation of momentum equation [10.28] and equate terms in each of the the unit vector directions to find the eastward component of the momentum equations,

$$\frac{Du}{Dt} - \frac{uv\tan\phi}{a} + \frac{uw}{a} = -\frac{1}{\rho}\frac{\partial P}{\partial x} + 2\Omega v\sin\phi - 2\Omega w\cos\phi + F_{\mathrm{f}x}, \tag{10.33}$$

the northward component of the momentum equations,

$$\frac{Dv}{Dt} + \frac{u^2\tan\phi}{a} + \frac{uw}{a} = -\frac{1}{\rho}\frac{\partial P}{\partial y} - 2\Omega u\sin\phi + F_{\mathrm{f}y}, \tag{10.34}$$

and the vertical component of the momentum equations,

$$\frac{Dw}{Dt} - \frac{u^2+v^2}{a} = -\frac{1}{\rho}\frac{\partial P}{\partial z} - \mathrm{g}_{\mathrm{eff}} + 2\Omega u\cos\phi + F_{\mathrm{f}z}. \tag{10.35}$$

10.4.2 The Conservation of Mass

The conservation of mass is often referred to as the continuity equation. We consider an infinitesimal volume and ask: what is the rate of increase of mass per unit volume $\partial\rho/\partial t$ of the mass inside the volume (i.e., the density change), due to mass flowing in and out of the volume? We start with mass flowing into the volume along the x direction through an area $\delta y\delta z$ (a face of a cube with sides δx, δy, δz). The mass flow rate (in units of kg s^{-1}) into the volume along the x direction through the area $\delta y\delta z$ is

$$\left(\rho u - \frac{\partial(\rho u)}{\partial x}\frac{\delta x}{2}\right)\delta y\delta z, \tag{10.36}$$

and the mass flow rate (in units of kg s^{-1}) out of the volume along the x direction through the area $\delta y \delta z$ is

$$\left(\rho u + \frac{\partial(\rho u)}{\partial x}\frac{\delta x}{2}\right)\delta y \delta z. \tag{10.37}$$

Subtracting the above two equations, we find the net rate of mass flow through the x-z sides of the volume to be

$$-\frac{\partial(\rho u)}{\partial x}\delta x \delta y \delta z. \tag{10.38}$$

Taking similar expressions for the mass flow along the y and z directions, we have for the net rate of mass flow into and out of the volume

$$\frac{\partial \rho}{\partial t}\delta x \delta y \delta z = -\left[\frac{\partial(\rho u)}{\partial x} + \frac{\partial(\rho u)}{\partial y} + \frac{\partial(\rho u)}{\partial z}\right]\delta x \delta y \delta z. \tag{10.39}$$

We can rewrite the above equation as

$$\boxed{\frac{\partial \rho}{\partial t} + \nabla \cdot (\rho \mathbf{U}) = 0.} \tag{10.40}$$

This is called the mass divergence form of the continuity equation.

10.4.3 The Conservation of Energy

The law of conservation of energy is used to relate the temperature of the atmosphere to heat sources and sinks. For example, with atmospheric dynamics in a differential layer with thickness Δz there is a net radiative flux

$$\Delta F(z) = F(z) - F(z + \Delta z), \tag{10.41}$$

where the absorbed radiation heats the layer. We may use similar arguments to those preceding the estimate of the radiative timescale (equation [10.1]) to express the layer heating by a rate of temperature change

$$\frac{\partial T}{\partial t} = -\frac{1}{\rho c_p}\frac{\Delta F(z)}{\Delta z}. \tag{10.42}$$

See, e.g., [9] for more details.

10.4.4 The Ideal Gas Law

The ideal gas law remains the same as before and is included here for completeness.

$$P = \rho R_{\mathrm{s}} T, \tag{10.43}$$

where R_{s} is the specific gas constant.

10.4.5 Models and the System of Conservation Equations

Atmospheric circulation models are based on the above six fundamental conservation equations: the conservation of mass, conservation of momentum (one equation for each dimension), conservation of energy, and the ideal gas law as the equation

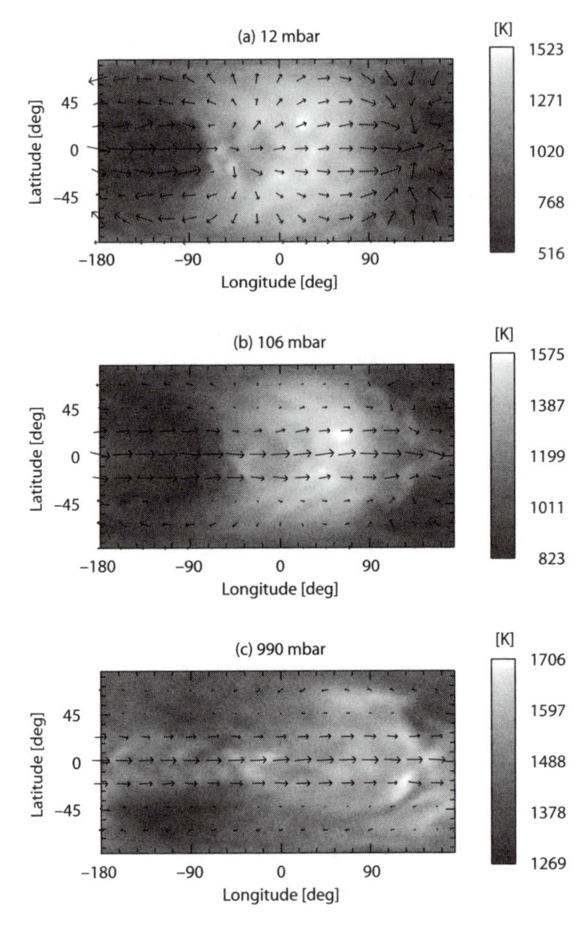

Figure 10.3 Simulated temperature of the tidally locked hot Jupiter HD 209458b. The arrows represent winds. From top to bottom are three different isobars: 1.5, 220, and 19.6 mbar. The substellar point is at (0,0) in longitude and latitude. Adapted from [8].

of state. The system of equations is not closed because of the unknown terms for friction and for the heating rate.

Many variations of conservation equations are used for atmospheric circulation models. Some researchers use the full set of conservation equations. Others take the traditional planetary atmospheres approach resulting from decades of study: the primitive equations. The primitive equations replace the vertical momentum equation with local hydrostatic balance, thereby dropping the vertical acceleration, advection, Coriolis, and metric terms that are generally expected to be less important for the global-scale circulation, such that energy is still conserved. For an example of the primitive equations in a general circulation model as applied to a hot Jupiter exoplanet see Figure 10.3. The term "primitive" refers to the full set of basic equations, before simplification by a suite of approximations. A different modeling

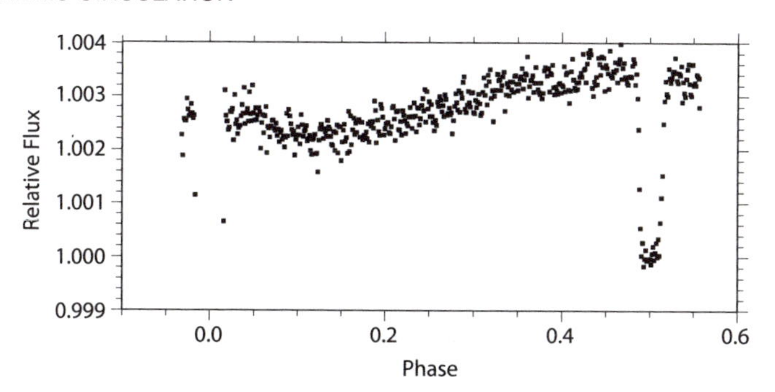

Figure 10.4 Thermal phase curve of HD 189733b. This is a zoomed-in section of Figure 1.4. Adapted from [10].

approach is to use a 2D version of the conservation equations, the shallow water equations. Because of the short timescales involved for radiation transport, atmospheric circulation models traditionally use somewhat elementary radiative transfer schemes.

The primitive equations or even more approximate forms are not easy to implement. Many institutions have developed their own General Circulation Models for public use. Many textbooks spend several chapters continuing the derivation of the conservation equations into different vertical coordinate systems and making many further approximations to make the equations more usable [e.g., 4, 7].

10.5 CONNECTION TO OBSERVATIONS

It is natural to ask what observations of exoplanets are directly connected to atmospheric circulation. At present, these are thermal phase curves of tidally locked hot exoplanets. The best example we have is shown in Figure 10.4, *Spitzer Space Telescope* 8 μm photometry of HD 189733b taken over 30 hours—half of the planet's 2.5-day-period orbit. We assume that at 8 μm the thermal variation of the combined planet and star flux is entirely due to the planet. It is fair to say, then, that the variation in Figure 10.4 corresponds to about 20% variation in planet temperature (from a brightness temperature of 1212 to 973 K) [10]. In contrast, other transiting exoplanets appear to have massive day-night temperature contrasts, up to 1000 K or even higher (e.g., [11]). The crude map shown in Figure 10.5 shows one possible map of HD 189733's 8 μm surface temperature. This map is misleading, because there is no real latitudinal information.

A second connection of atmospheric dynamics to observations is related to the study of Earth as an exoplanet. A pale blue dot observed from afar, Earth is actually the most variable object in our solar system at visible wavelengths. This is due to water clouds and their high albedos in contrast to the dark oceans. Despite the apparent variability of clouds, large-scale patterns persist long enough so that we can determine the rotation rate of Earth as viewed from afar. By binning data to the

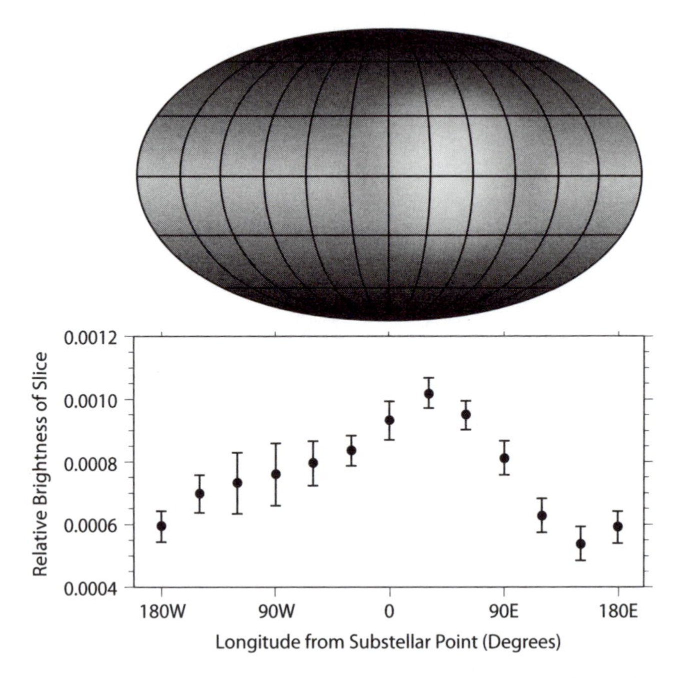

Figure 10.5 Brightness estimates for 12 longitudinal strips on the surface of HD 189733b. a) data are shown as an example a color map. b) data are shown in graphical form. Adapted from [10].

rotation period, we can see variability and attribute the variability to the presence of clouds [12, 13].

One intriguing and promising avenue to understanding what the data can really tell us about exoplanet atmosphere observations may come from inversion of the observed thermal phase functions [14]. This is akin to first trying to understand where hot or cold spots are in the exoplanet atmosphere and next trying to understand the degeneracy in terms of latitude and longitude (and with spectral observations, altitude). Then one could attempt to attribute physical mechanisms for the specific variations. The method could also work for inverting the scattered light phase functions of Earth-like exoplanets, whereby one could infer, very crudely, the presence of continents.

In the future, with the launch of NASA's *James Webb Space Telescope*, spectra as a function of orbital phase will give us a better picture of thermal phase variation. In particular, high signal-to-noise data spectra will not only give us the longitudinally averaged thermal flux but also enable us to understand the flux as a function of altitude (see Chapter 6).

10.6 SUMMARY

Atmospheric circulation is a fundamental topic for planetary atmospheres because it plays such a significant role in how exoplanets look from afar. For many planets, atmospheric circulation acts to minimize horizontal temperature gradients, including equator-to-pole temperature gradients. For solar system planets, the temperature gradients are so small as to not be observable in the hemispherically averaged spectra (except notably in a hemispherically integrated spectrum of Earth centered on Earth's cold poles). For such planets, the 1D average temperature-pressure structure calculations described in Chapter 9 can be used to understand the temperature from observed spectra.

Alternatively, there are exoplanets where atmospheric circulation sets up strong temperature gradients. Evidence comes from thermal phase curve measurements of a few hot Jupiters. For some of these tidally locked planet cores, even the thick mobile atmosphere is not able to circulate the absorbed stellar energy. Instead, hot and cold spots and even hot and cold sides of the planet are maintained. Atmospheric dynamics also plays a role in studies of Earth as an exoplnet—large-scale cloud patterns and their variability may be detectable in the hemispherically integrated signal of visible-wavelength scattered light.

REFERENCES

For further reading

An excellent introduction to atmospheric circulation:

- Marshall, J., and Plumb, A. 2008. *Atmosphere, Ocean, and Climate Dynamics: An Introductory Text* (London: Elsevier Academic Press).

Two fundamental references on atmospheric circulation:

- Vallis, G. K. 2006. *Atmospheric and Oceanic Fluid Dynamics* (Cambridge: Cambridge University Press).
- Holton, J. R. 1992. *An Introduction to Dynamic Meteorology* (3rd ed; San Diego: Academic Press).

The most thorough reference focused on exoplanet atmospheric circulation is:

- Showman, A. P., Cho, J. Y.-K., and Menou, K. 2010. "Atmospheric Circulation of Exoplanets," in *EXOPLANETS*, ed. S. Seager (Tucson: University of Arizona Press)

References for this chapter

1. Ingersoll, A. P. 1990. "Atmospheric Dynamics of the Outer Planets," Science 248, 308–315.

2. Showman, A. P., Menou, K., and Cho, J. Y.-K. 2008. "Atmospheric Circulation of Hot Jupiters: A Review of Current Understanding," in *Extreme Solar Systems* eds. D. Fischer, F. A. Rasio, S. E. Thorsett, and A. Wolszczan, Astronomical Society of the Pacific Conference Series 398, 419–441.

3. Seager, S., Richardson, L. J., Hansen, B. M. S., Menou, K., Cho, J. Y.-K., and Deming, D. 2005. "On the Dayside Thermal Emission of Hot Jupiters." Astrophys. J. 632, 1122–1131.

4. Holton, J. R. 1992. *An Introduction to Dynamic Meteorology* (3rd Ed; San Diego: Academic Press).

5. Rhines, P. B. 1975. "Waves and Turbulence on a Beta-Plane." J. Fluid Mech. 69, 417–433.

6. Menou, K., Cho, J. Y.-K., Seager, S., and Hansen, B. M. S. 2003. "Weather Variability of Close-in Extrasolar Giant Planets." Astrophys. J. 587, L113–116.

7. Pedlosky, J. 1987. *Geophysical Fluid Dynamics* (2nd ed; New York: Springer).

8. Cooper, C. S., and Showman, A. P. 2005. "Dynamic Meteorology at the Photosphere of HD 209458b." Astrophys. J. 629, L45–48.

9. Liou, K. N. 2002. *An Introduction to Atmospheric Radiation* (London: Academic Press).

10. Knutson, H. A., et al. 2007. "A Map of the Day-Night Contrast of the Extrasolar Planet HD 189733b." Nature 447, 183–186.

11. Harrington, J., Hansen, B. M., Luszcz, S. H., Seager, S., Deming, D., Menou, K., Cho, J. Y.-K., and Richardson, L. J. 2006. "The Phase-Dependent Infrared Brightness of the Extrasolar Planet Upsilon Andromedae b." Science 314, 623–626.

12. Ford, E. B., Seager, S., and Turner, E. L. 2001. "Characterization of Extrasolar Terrestrial Planets from Diurnal Photometric Variability." Nature 412, 885–887.

13. Palle, E., Ford, E. B., Seager, S., Montanes-Rodriguez, P., and Vazquez, M. 2008. "Identifying the Rotation Rate and the Presence of Dynamic Weather on Extrasolar Earth-Like Planets from Photometric Observations." Astrophys. J. 676, 1319–1329.

14. Cowan, N. B., and Agol, E. 2008. "Inverting Phase Functions to Map Exoplanets." Astrophys. J. 678, L129–132.

EXERCISES

1. Derive the radiative timescale in equation [10.1].

2. How many bands should Earth have, using equation [10.21]?

3. Derive equation [10.13] from equation [10.12]. Use the definition of $D\mathbf{U}/Dt$ and explain and use the vector identity

$$\boldsymbol{\Omega} \times (\boldsymbol{\Omega} \times \mathbf{r}) = \boldsymbol{\Omega} \times (\boldsymbol{\Omega} \times \mathbf{R}) = -\Omega^2 \mathbf{R}. \quad (10.44)$$

   Here $\boldsymbol{\Omega}$ is taken to be a constant and $\mathbf{R}$ is a vector perpendicular to the axis of rotation. The magnitude of $\mathbf{R}$ is equal to the distance to the axis of rotation.

4. Show that the relation between the total rate of change of a variable is related to the local derivative at a fixed point is

$$\frac{\partial \mathbf{A}}{\partial t} = \frac{D\mathbf{A}}{Dt} - \mathbf{U} \cdot \nabla \mathbf{A}, \quad (10.45)$$

   where $\mathbf{U}$ is velocity.

5. A longitudinally averaged thermal flux of HD 189733b at 8μm is shown in the bottom panel of Figure 10.5, and a map is shown in the top panel. Describe why the map is not unique. Sketch two other representations of the map that also satisfy the actual measured data.

Chapter Eleven

Atmospheric Biosignatures

11.1 INTRODUCTION

A major goal of the study of exoplanet atmospheres is to look for biosignature gases that may indicate the presence of life on a distant exoplanet. In fact, one of the driving factors for the entire field of exoplanet research is often described as the aim to discover and characterize a true Earth twin, with the express hopes of finding signs of life in the atmosphere. In this chapter we deviate from the quantitative formality in the rest of book, in order to close with a discussion of a very exciting avenue for the application of exoplanet atmosphere studies.

11.2 EARTH'S BIOSIGNATURES

Earth's atmosphere is a natural starting point when we consider planets that may have the right conditions for life or planets that have signs of life in their atmospheres. Earth from afar, with any reasonably sized telescope, would appear only as a point of light, without spatial resolution of surface features. We have real atmosphere spectra of the hemispherically integrated Earth (Figure 1.2) by way of Earthshine measurements at visible and near-IR wavelengths and from spacecraft that turn to look at Earth.

Earth has several strong spectral features that are uniquely related to the existence of life or habitability. The gas oxygen (O_2) makes up 21% of Earth's atmosphere by volume, yet O_2 is highly reactive and therefore will remain in significant quantities in the atmosphere only if it is continually produced. On Earth plants and photosynthetic bacteria generate oxygen as a metabolic by-product; there are no abiotic continuous sources of large quantities of O_2. Ozone (O_3) is a photolytic product of O_2, generated after O_2 is split up by the Sun's UV radiation. Oxygen and ozone are Earth's two most robust biosignature gases. Nitrous oxide (N_2O) is a second gas produced by life—albeit in small quantities—during microbial oxidation-reduction reactions. For a detailed discussion and simulated atmospheric spectra, see [1].

Other spectral features, although not biosignatures because they do not reveal direct information about life or habitability, can nonetheless provide significant information about the planet. These include CO_2 (which is indicative of a terrestrial atmosphere and has an extremely strong midinfrared spectral feature) and CH_4 (which has both biotic and abiotic origins). Thus detection of several of these features would provide credible evidence for the existence of life as we know it.

In addition to atmospheric biosignatures, the Earth has one very strong and very intriguing biosignature on its surface: vegetation [2, 3, 4, and references therein]. The reflection spectrum of photosynthetic vegetation has a dramatic sudden rise in albedo around 750 nm by almost an order of magnitude. Vegetation has evolved this strong reflection feature, known as the "red edge," as a cooling mechanism to prevent overheating which would cause chlorophyll to degrade. On Earth this signature is probably reduced to a few percent because of clouds. But such a spectral surface feature could be much stronger on a planet with a lower cloud cover fraction. Recall that any observations of Earth-like extrasolar planets will not be able to spatially resolve the surface. A surface biosignature could be distinguished from an atmospheric signature by time variation; as the continents, or different concentrations of the surface biomarker, rotate in and out of view the spectral signal will change correspondingly.

Although not a biosignature, photometric variation of Earth-like exoplanets may be indicative of clouds, which in turn are suggestive of liquid water oceans. Liquid water is considered a major habitability indicator. Earth viewed from afar would vary in brightness with time, due to the brightness contrast of cloud, land, and oceans. As Earth rotates and continents come in and out of view, the total amount of reflected sunlight will change due to the high albedo contrast of different components of Earth's surface ($<$ 10% for ocean, $>$30%–40% for land, $>$ 60% for snow and some types of ice). In the absence of clouds this variation could be an easily detectable factor of a few. With clouds the variation is muted to 10% to 20% [5]. From continuous observations of Earth over a few-month period, Earth's rotation rate could be extracted, weather identified, and the presence of continents inferred [6]. A hypothetical distant observer of Earth could measure Earth's rotation rate. This is surprising and means that, despite Earth's dynamic weather patterns, Earth has a relatively stable signature of cloud patterns. These cloud patterns arise in part because of Earth's continental arrangement and ocean currents. Beyond detection of Earth's rotation rate, therefore, deviations from the periodic photometric signal, would be indicative to hypothetical distant observers that active weather is present on Earth, in turn indicating the presence of liquid water oceans.

11.3 THE IDEAL BIOSIGNATURE GAS

The ideal atmospheric biosignature gas would have the following characteristics. It would

1. not exist naturally in the planetary atmosphere at ambient temperatures and pressures;

2. not be created by geophysical processes;

3. not be produced by photochemistry; and

4. have a strong spectral signature.

Indeed, Earth's most robust biosignature gas O_2 satisfies all four of the above criteria and N_2O satisfies the first three. Even though O_2 and N_2O are also generated by photochemistry, the amounts are minuscule.

It is interesting to consider the list of gases emitted by microbial life on Earth. A partial list includes O_2, H_2, CO_2, N_2, N_2O, NO, NO_2, H_2S, SO_2, and CH_4 (e.g., [7]). Let us go through this list and consider the feasibility of each as a biosignature. We have already briefly described Earth's biosignature gases O_2, N_2O, and CH_4. On Earth, the gas N_2 makes up 78% of our atmosphere and is therefore not a useful biosignature. Furthermore, as a homonuclear molecule N_2 has no rotational-vibrational transitions and hence no spectral signature at visible and infrared wavelengths. Like N_2, CO_2 is already present in Earth's atmosphere, making up about 0.035%. For Venus and Mars (the only other solar system terrestrial planets with an atmosphere) CO_2 makes up more than 97% of the atmosphere. As such, CO_2 is considered a major planetary atmosphere gas that results from planet formation and evolution, and hence is not a useful biosignature. The gases H_2, H_2S, SO_2, NO, and NO_2. are produced on Earth either by volcanoes or by photochemistry, making them nonunique biosignatures. On Earth these gases are produced in tiny quantities (both biotically and abiotically) and hence also lack any detectable spectral signature for a remote observer.

Many exoplanets may simply not have a unique biosignature like Earth's O_2 and N_2O. In our preparation for the search for exoplanet biosignatures, we must aim to consider the more common gases produced both as metabolic by-products and by either geophysical or photochemical processes. To understand their potential as biosignatures, we would have to model exoplanet environment scenarios to determine under which cases the common gases would be much more abundant than can reasonably be produced by geologic or photochemical processes. In other words, absent a unique biosignature such as oxygen, the main criterion for a biosignature is that the gas exists in such great quantities that its presence in a planetary atmosphere is well above the amounts that could be produced by any abiotic processes. In such a scenario we may be relatively certain but never 100% positive that we have found signs of life on another planet. Alternatively, we may also be guided toward identifying biosignatures by a purely observational approach. That is one of the early ideas in biosignature research: to find a planet atmosphere that is out of chemical equilibrium, especially in redox disequilibrium [8,9].

11.4 PROSPECTS

11.4.1 Discovery and Characterization Prospects for a True Earth Analog

Without question the holy grail of the research field of exoplanets is the discovery of a true Earth analog: an Earth-size, Earth-mass planet in an Earth-like orbit about a sunlike star. Discovery of an Earth analog is a massive challenge for each of the different exoplanet discovery techniques (Figure 1.5), because Earth is so much smaller (1/100 in radius), so much less massive ($1/10^6$), and so much fainter (10^7 for mid-IR wavelengths to 10^{10} for visible wavelengths) than the Sun.

In order to observe a planet under such low planet-star flux contrasts, at visible wavelengths the diffracted light from the star must be suppressed by 10 billion times. A collection of related ideas to suppress the starlight are called "coronagraphs," a term that originated with the artificial blocking out of sunlight to observe the Sun's corona. Novel-shaped apertures, pupils, and pupil masks to suppress some or all of the diffracted star light have been developed [10 and references therein]. A very promising idea is that of a novel-shaped external occulter that would fly tens of thousands of kilometers from the telescope in order to suppress the diffracted starlight [11].

Exoplanet discovery techniques other than direct imaging, (shown in Figure 1.5), may have an easier time in discovering an Earth-mass or Earth-size planet. These other methods include space-based astrometry (see [12] and references therein) or potentially ground-based radial velocity with the new development of the astrofrequency comb spectrograph [e.g., 13]. NASA's *Kepler Space Telescope* is using the transit technique to take a census of Earth-size, Earth-like transiting planets orbiting sunlike stars.

Despite the near-future prospects for finding Earth-mass or Earth-size planets in Earth-like orbits, we emphasize that discovery of Earth-size or Earth-mass planets is not the same as identifying a habitable planet. Venus and Earth are both about the same size and same mass—and would appear the same to an astrometry, radial-velocity, or transit discovery. Yet Venus is completely hostile to life due to the strong greenhouse effect and resulting high surface temperatures (over 700 K), while Earth has the right surface temperature for liquid water oceans and is teeming with life. This is why, in the search for habitable planets, we must hold on to the dream of a direct-imaging space-based telescope capable of blocking out the starlight.

11.4.2 Discovery and Characterization Prospects for a Super Earth Orbiting an M star

The discovery and spectral characterization of a true Earth analog are immensely challenging, making them at present many years off in the future. Yet a different kind of habitable planet is within reach—if only we are willing to extend our definition of Earth. The extension is a big Earth orbiting close to a small star.

All life on Earth requires liquid water and so a natural requirement in the search for habitable exoplanets is a planet with the right surface temperature to have liquid water. Terrestrial-like planets are heated externally by the host star, so that a star's "habitable zone" is based on distance from the host star. Small stars have a habitable zone much closer to the star as compared to sunlike stars, owing to the small star's much lower energy output than the Sun.

The chance of discovering a super Earth transiting in the habitable zone of a low-mass star in the immediate future is huge. Observational selection effects favor their discovery in almost every way (see Figure 11.1). The magnitude of the planet transit signature is related to the planet-to-star area ratio. Low-mass stars can be 2–10 times smaller than a sunlike star, improving the transit signal from about 1/100,000

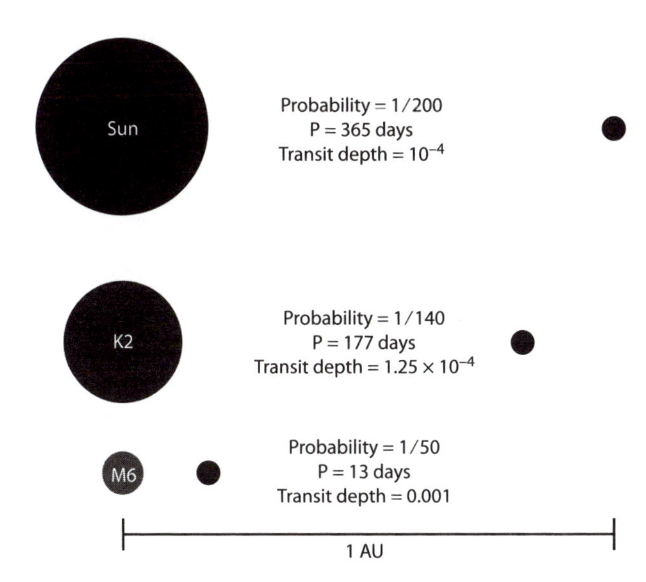

Figure 11.1 A schematic of transiting planets orbiting normal stars. The smaller the star, the closer the habitable zone is to the star, and the easier detection is.

for an Earth transiting a Sun, to 1/25,000 or 1/1000 for the same-sized planet. A planet's equilibrium temperature scales as $T_{\rm eq} \sim T_*(R_*/a)^{1/2}$ (equation [3.9]), where a is the planet's semimajor axis. The temperature of low-mass stars is about 3500–2800 K (for the star respectively 2 and 10 times smaller than the Sun). The habitable zone of a low-mass star would therefore be 4.5–42 times closer to the star compared to the Sun's habitable zone. To measure a planet mass, the radial velocity amplitude K scales as $K \sim (aM_*)^{-1/2}$, and the low-mass star masses are 0.4–0.06 times that of the Sun. It is therefore about 3–30 times more favorable to obtain a planet mass for an Earth-mass planet orbiting a low-mass star compared to an Earth-Sun analog. The transit probability scales as R_*/a. The probability for a planet to transit in a low-mass star's habitable zone is about 2.3%–5%, much higher as compared to the low 0.5% probability of an Earth-Sun analog transit. Finally, from Kepler's Third Law a planet's period P scales as $P \sim a^{3/2}/M^{1/2}$, meaning that the period of a planet in the habitable zone of a low-mass star is 7–90 times shorter than the Earth's 1 year period, and the planet transit can be observed often enough to build up a signal (as compared to an Earth analog's once-a-year transit). A super Earth larger than Earth (and up to about $10M_\oplus$ and $2R_\oplus$) is even easier to detect by its larger transit signal and mass signature than those of Earth. For further discussion on M stars and planets, see [14].

We anticipate the discovery of a handful of the rare but highly valuable transiting super Earths in the habitable zones of low-mass stars. With such prize targets, astronomers will strive to observe the transiting super Earth atmospheres in the same way that we are currently observing transiting hot Jupiters orbiting sunlike stars. The observations will be challenging, due to the thinner atmospheres on rocky planets compared to the puffy atmospheres of hot Jupiter (gas giant) planets. Also, the stars could be faint if Earths in a low-mass-star habitable zone are rare. Fortunately, NASA's *James Webb Space Telescope* (JWST), scheduled for launch in 2014, is capable of observing super Earths transiting low-mass stars [e.g., 15]. We need to be patient with the tens to hundreds of hours of *JWST* time with the concomitant complex scheduling to cover many periodic transits in order to build up a decent signal of the atmospheric spectrum.

There is an exciting sense of anticipation in observing and studying super Earth atmospheres, simply because we don't know quite what their diversity will be. This is in contrast to Jupiter and the other solar system gas giants, which have a "primitive" atmosphere. That is, Jupiter has retained the gases it formed with, and these gases represent the composition of the Sun. The super Earth atmospheres, on the other hand, could have a wide range of possibilities for the atmosphere mass and composition, right down to the amount of hydrogen in the atmosphere. Hydrogen is a light gas and escapes from small, warm terrestrial planets. Earth and the other solar system terrestrial planets have all lost whatever hydrogen they may have started with. A hydrogen-rich atmosphere is very intriguing because it might have very different biosignatures from those of an oxidized atmosphere.

11.5 SUMMARY

The research area of exoplanet biosignatures is so far very terracentric. There is fertile ground for new research in the context of exoplanet atmospheres very different from Earth's. Overall, the field of exoplanet atmospheres is an exciting and rapidly growing area of scientific research. The *Spitzer Space Telescope* has revolutionized observations of transiting hot Jupiter atmospheres (see examples in Chapter 6). We have a lot to look forward to in the future with the launch of *JWST* and its capability of observing transiting super Earth atmospheres for super Earths in the habitable zones of M stars. The single most exciting possibility is the construction and launch of an external occulter to work in conjunction with *JWST*. Such a combination could find and directly image true Earth analogs around a few dozen nearby stars—and would succeed if Earths are very common. A related possibility, in the more distant future, is the hope for the launch of a space telescope capable of discovery and direct imaging of a large number of true Earth analogs.

REFERENCES

For further reading

- Seager, S., and Meadows, V. 2010. "Biosignatures," in *EXOPLANETS*, ed. S. Seager (Tucscon: University of Arizona Press).

References for this chapter

1. Des Marais, D. J., et al. 2002. "Remote Sensing of Planetary Properties and Biosignatures on Extrasolar Terrestrial Planets." Astrobiology 2, 153–181.
2. Woolf, N. J., Smith, P. S., Traub, W. A., and Jucks, K. W. 2002. "The Spectrum of Earthshine: A Pale Blue Dot Observed from the Ground." Astrophys. J. 574, 430–433.
3. Seager, S., Turner, E. L., Schafer, J., and Ford, E. B. 2005. "Vegetation's Red Edge: A Possible Spectroscopic Biosignature of Extraterrestrial Plants." Astrobiology 5, 372–390.
4. Montanes-Rodriguez, P., Palle, E., Goode, P. R., and Martin-Torres, F. J. 2006. "Vegetation Signature in the Observed Globally Integrated Spectrum of Earth Considering Simultaneous Cloud Data: Applications for Extrasolar Planets." Astrophys. J. 651, 544–52.
5. Ford, E. B., Seager, S., and Turner, E. L. 2001. "Characterization of Extrasolar Terrestrial Planets from Diurnal Photometric Variability." Nature 412, 885–887.
6. Palle, E., Ford, E. B., Seager, S., Montanes-Rodriguez, P., and Vazquez, M. 2008. "Identifying the Rotation Rate and the Presence of Dynamic Weather on Extrasolar Earth-Like Planets from Photometric Observations." Astrophys. J. 676, 1319–1329.
7. Madigan, M. M, Martinko, J., and Parker, J. 2001. *Brock Biology of Microorganisms* (Boston: Prentice Hall).
8. Lovelock, J. 1965. "A Physical Basis for Life Detection Experiments." Nature. 207, 568–570.
9. Lederberg, J. 1965. "Signs of Life—Criterion-System of Exobiology." Nature 207, 9–13.
10. Trauger, J. T., and Traub. W. A. 2007. "A Laboratory Demonstration of the Capability to Image an Earth-Like Extrasolar Planet." Nature 446, 771–773.
11. Cash, W. 2006. "Detection of Earth-Like Planets Around Nearby Stars Using a Petal-Shaped Occulter." Nature 442, 51–53.
12. Shao, M., and Nemati, B. 2009. "Sub-Microarcsecond Astrometry with SIM-Lite: A Testbed-Based Performance Assessment." Pub. Astron. Soc. Pac. 121, 41–44.
13. Li, C.-H. et al. 2008. "A Laser Frequency Comb that Enables Radial Velocity Measurements with a Precision of 1 cm s^{-1}." Nature, 452, 610-612.
14. Tarter. J et al. 2007. "A Reappraisal of Planets around M Dwarf Stars." Astrobiology 7, 30–65.

15. Deming et al. 2009. "Discovery and Characterization of Transiting Super Earths Using an All-Sky Transit Survey and Follow-Up by the *James Webb Space Telescope*." Pub. Astron. Soc. Pac. 121, 952–967.

Appendix A

Planetary Data

Table A.1 Physical properties of solar system planets.

Planet	Mass ($M_\oplus$)	Radius ($R_\oplus$)	a (AU)	e	$T_{\rm eff}(K)$	Oblate-ness
Mercury	0.055	0.383	0.387	0.206	—	0.0
Venus	0.815	0.949	0.723	0.007	~ 230	0.0
Earth	1.000	1.000	1.000	0.017	~ 255	0.003
Mars	0.107	0.534	1.524	0.093	~ 212	0.006
Jupiter	317.82	11.209	5.203	0.048	124.4 ± 0.3	0.065
Saturn	95.161	9.449	9.537	0.054	95.0 ± 0.4	0.098
Uranus	14.371	4.007	19.191	0.047	59.1 ± 0.3	0.023
Neptune	17.147	3.883	30.069	0.009	59.3 ± 0.8	0.017

From left to right the columns are planet, mass, equatorial radius, semimajor axis (a), effective temperature and oblateness = $\frac{R_{\rm e}-R_{\rm p}}{R_{\rm p}}$. Values (rounded off) from [1].

Table A.2 Mixing ratios of major and spectrally significant gases in solar system terrestrial planet atmospheres.

Planet	N_2	Ar	CO_2	O_2	O_3	H_2O
Earth	78.084	< 0.934	350 ppm	20.946	$\sim$10–100 ppb	$< 4^*$
Venus	3.5 ± 0.8	–	96.5 ± 0.8	–	–	–
Mars	2.7	1.6	95.32	–	–	–

Values are in % unless otherwise noted. By major we mean a mixing ratio $> 1\%$. By spectrally significant we mean a gas that contributes spectral features at visible through infrared wavelenths that would be detectable for low S/N ($\sim$ 10) low spectral resolution ($\sim$ 70) data. A dash means the gas is not major nor spectrally significant. The superscript * means variable. Data from [2] and see references therein.

Table A.3 Mixing ratios of the three most abundant gases in solar system giant planet atmospheres.

Planet	H_2	He	CH_4
Jupiter	86.2 ± 2.6	13.6 ± 2.6	0.21 ± 0.04
Saturn	96.3 ± 2.4	3.25 ± 2.4	0.45 ± 2.4
Uranus	$\sim 82.5 \pm 3.3$	15.2 ± 3.3	~ 2.3
Neptune	$\sim 80 \pm 3.2$	19.0 ± 3.2	$\sim$ 1–2

Values are in %. Data from [2] and see references therein.

Table A.4 Exosphere values for Earth, Venus, and Mars.

Planet	r_c (km)	T_c (K)
Earth	500	1000
Venus	200	275
Mars	250	300

Values from [4].

Table A.5 Star characteristics for planet-star flux ratio estimates in Chapter 3.

Spectral Type	$M/M_\odot$	$R/R_\odot$	$\log \overline{\rho}/\overline{\rho}_\odot$	$T_{\rm eff}$ (K)
O5	60	12	-1.5	42,000
B0	17.5	7.4	-1.4	30,000
B5	5.9	3.9	-1.00	15,200
B8	3.8	3.0	-0.85	11,400
A0	2.9	2.4	-0.7	9760
A5	2.0	1.7	-0.4	8180
F0	1.6	1.5	-0.3	7300
F5	1.4	1.3	-0.2	6650
G0	1.05	1.1	-0.1	5940
G5	0.92	0.92	-0.1	5560
K0	0.79	0.85	0.1	4550
K5	0.67	0.72	0.25	3990
M0	0.51	0.60	0.35	3620
M2	0.40	0.50	0.8	3370
M5	0.21	0.27	1.0	2880
M8	0.06	0.1	1.2	2500

Values from [1].

Table A.6 Saturation vapor pressure and latent heat values.

Component	ln (C) [C in bars]	L_0 [J g^{-1}]	$\Delta\alpha$ [J g^{-1} K^{-1}]	$\Delta\beta/2$ [J g^{-1} K^{-2}]
H_2O	25.096	3148.2	–	-8.7 $\times 10^{-3}$
CO_2	26.100	639.6	–	-1.7 $\times 10^{-3}$
NH_3	27.863	2016	-0.888	–
NH_4SH	75.678	2915.7	-1.760	7.8 $\times 10^{-4}$
CH_4	1.627	553.1	1.002	-4.1 $\times 10^{-3}$
Fe	1.894	7097	–	–
$MgSiO_3$	11.554	4877.5	–	–

To be used with equation [4.40]. Adapted from [3] and see references therein.

REFERENCES

1. Cox, A.N.. *Allen's Astrophysical Quantities* (4th ed; New York: Springer).

2. Lodders, K., and Fegley, B. 1998. *The Planetary Scientist's Companion* (Oxford: Oxford University Press).

3. Sanchez-Lavega, A., Perez-Hoyos, S., and Hueso R. 2004. "Clouds in planetary atmospheres: A useful application of the Clausius-Clapeyron equation." Am. J. Phys. 72, 767–774.

4. Hunten, D. M. 1982. "Thermal and Nonthermal Escape Mechanisms for Terrestrial Bodies." Planet. Space Sci. 30, 773–783.

Index